Transportation Planning Applications 2018

Transportation Research Record: Journal of the Transportation Research Board Volume 2672, Issue 46

ISBN: 978-1-5108-8252-2

Printed from e-media with permission by:

Curran Associates, Inc.
57 Morehouse Lane
Red Hook, NY 12571

CURRAN ASSOCIATES INC.
proceedings
.com

Some format issues inherent in the e-media version may also appear in this print version.

Published by Sage Publications USA

ISBN (Print) 978-1-5108-8252-2
ISBN 2018 Printed Set (All Issues) 978-1-5108-7735-1

Printed by Curran Associates, Inc. (2019)

Additional copies of this publication are available from:

Curran Associates, Inc.
57 Morehouse Lane
Red Hook, NY 12571 USA
Phone: 845-758-0400
Fax: 845-758-2633
Email: curran@proceedings.com
Web: www.proceedings.com

Volume 2672 Number 46 December 2018

TRR
JOURNAL OF THE TRANSPORTATION RESEARCH BOARD

Contents

Transportation Planning Applications

Articles

Rough Roads Ahead 2: Economic Implications of Deteriorating Highway Conditions 1
Becky Knudson and Alex Bettinardi

How Autonomous Driving May Affect the Value of Travel Time Savings for Commuting 11
Felix Steck, Viktoriya Kolarova, Francisco Bahamonde-Birke, Stefan Trommer, and Barbara Lenz

Evaluating the Ability of Transit Direct Ridership Models to Forecast Medium-Term Ridership Changes: Evidence from San Francisco 21
Richard A. Mucci and Gregory D. Erhardt

A Model of Ridesourcing Demand Generation and Distribution 31
Patrícia S. Lavieri, Felipe F. Dias, Natalia Ruiz Juri, James Kuhr, and Chandra R. Bhat

Street Intersection Characteristics and Their Impacts on Perceived Bicycling Safety 41
Kailai Wang and Gulsah Akar

A German Passenger Car and Heavy Vehicle Stock Model: Towards an Autonomous Vehicle Fleet 55
Martin Hartmann and Peter Vortisch

The Pros and Cons of Using the Change in Destination Choice Logsums as a Practical Measure of User Benefits 64
Kristen Villanueva, Lisa Zorn, David Ory, and David Vautin

Using Google's Passive Data and Machine Learning for Origin-Destination Demand Estimation 73
Bhargava Sana, Joe Castiglione, Drew Cooper, and Dan Tischler

---

* Membership as of February 2019.

TRB

TRANSPORTATION RESEARCH BOARD

CORE SPONSORS

Transportation Departments of the 50 States and the District of Columbia

Federal Government

U.S. Department of Transportation
- Federal Aviation Administration (FAA)
- Federal Highway Administration
- Federal Motor Carrier Safety Administration (FMCSA)
- Federal Railroad Administration (FRA)
- Federal Transit Administration
- National Highway Traffic Safety Administration (NHTSA)
- Office of the Assistant Secretary of Transportation for Research and Technology (OST-R)

U.S. Air Force Civil Engineer Center (AFCEC)
U.S. Coast Guard (USCG)
U.S. Department of Energy
U.S. Department of the Interior (DOI)
U.S. Environmental Protection Agency (EPA)

Nongovernmental Organizations

American Public Transportation Association (APTA)
Association of American Railroads (AAR)
California Air Resources Board

The National Academies of

SCIENCES · ENGINEERING · MEDICINE

The **National Academy of Sciences** was established in 1863 by an Act of Congress, signed by President Lincoln, as a private, nongovernmental institution to advise the nation on issues related to science and technology. Members are elected by their peers for outstanding contributions to research. Dr. Marcia McNutt is president.

The **National Academy of Engineering** was established in 1964 under the charter of the National Academy of Sciences to bring the practices of engineering to advising the nation. Members are elected by their peers for extraordinary contributions to engineering. Dr. C. D. Mote, Jr., is president.

The **National Academy of Medicine** (formerly the Institute of Medicine) was established in 1970 under the charter of the National Academy of Sciences to advise the nation on medical and health issues. Members are elected by their peers for distinguished contributions to medicine and health. Dr. Victor J. Dzau is president.

The three Academies work together as the National Academies of Sciences, Engineering, and Medicine to provide independent, objective analysis and advice to the nation and conduct other activities to solve complex problems and inform public policy decisions. The Academies also encourage education and research, recognize outstanding contributions to knowledge, and increase public understanding in matters of science, engineering, and medicine.

Learn more about the National Academies of Sciences, Engineering, and Medicine at **www.national-academies.org.**

The **Transportation Research Board** is one of seven major programs of the National Academies of Sciences, Engineering, and Medicine. The mission of the Transportation Research Board is to increase the benefits that transportation contributes to society by providing leadership in transportation innovation and progress through research and information exchange, conducted within a setting that is objective, interdisciplinary, and multimodal. The Board's varied activities annually engage about 7,000 engineers, scientists, and other transportation researchers and practitioners from the public and private sectors and academia, all of whom contribute their expertise in the public interest. The program is supported by state transportation departments, federal agencies including the component administrations of the U.S. Department of Transportation, and other organizations and individuals interested in the development of transportation.

Learn more about the Transportation Research Board at **www.TRB.org**.

Article

Transportation Research Record
2018, Vol. 2672(46) 1–10

DOI: 10.1177/0361198118758053
journals.sagepub.com/home/trr
SAGE

Rough Roads Ahead 2: Economic Implications of Deteriorating Highway Conditions

Becky Knudson[1] and Alex Bettinardi[1]

Abstract

This study presents a high-level, strategic assessment of the long-range implications of Oregon's current forecast budget for highways. The current highway budget is insufficient to meet current state goals with respect to transportation system condition, requiring the Oregon Department of Transportation (ODOT) to prepare for strategic investment decisions. ODOT must conceive new ways to develop the most efficient use and management of facilities while accepting lower performance standards, shifted use of deteriorating facilities, and the need to allocate resources to priority locations.The Oregon Statewide Integrated Model was used to evaluate three hypothetical scenarios over a 20-year period, representing investment strategies implemented with budget increases and compared with the current 20-year forecast budget. The purpose of the analysis was to evaluate the potential impact of highway disinvestment on Oregon's economic competitiveness. The primary focus was on bridges, as over half of Oregon's bridges are nearing the end of their life cycle. Weight-restricted bridges significantly impact the cost of moving freight in Oregon, where over 70% of freight moves by truck. In addition, registered passenger vehicle records were used to estimate the impacts of deteriorating pavement conditions on Oregon highway users.

Over the last 10 years Oregon transportation planning analysis has put modeling tools to use evaluating strategic policy options prepared for the Oregon Transportation Plan (*1*), identifying implications on freight movement under different economic conditions (*2*), and exploring economic implications of a major seismic event (*3*). This study evaluates the potential impacts of delaying key investments in Oregon's highway system, and builds on the scenario analysis approach prepared in a previous study, Rough Roads Ahead: The Cost of Poor Highway Conditions to Oregon's Economy (*4*).

The Oregon economy is currently very strong. The state has been adding jobs faster than the U.S. since 2013, recently adding more jobs in 2015 than over any 12-month period since 1990, despite the unemployment rate being the lowest in 40 years (*5*). Oregon's export-dependent economy fluctuates with business cycles (*6*).Traded industries rely on transportation to ship goods to customers outside the state, which generates revenue and creates jobs. A Portland study (*7*) estimated 1:5 jobs in the state of Oregon is transportation reliant or transportation related.

Transportation system deterioration occurs gradually over time. Recently, expenditures have shifted to preservation and maintenance as facilities approach the end of their design life and the cost to maintain and replace these large assets becomes a larger proportion of the required work. Present funding levels are insufficient to maintain current conditions. If funding remains the same, infrastructure will fall below the standards defined by the U.S. Department of Transportation (U.S. DOT) as "minimum tolerable conditions" for pavements and bridges. This pattern holds true across the nation, especially so for Oregon bridges, because the majority of bridges on state highways have reached or are approaching the end of their expected life.

Literature Review

A number of states have published reports utilizing funding scenarios, including studies representing current funding levels for Wisconsin (*8*), and Indiana (*9*), evaluating deteriorating pavements and bridges in terms of overall percentages, but have not included economic impacts. A Washington

[1]Oregon Department of Transportation, Salem, OR

Corresponding Author:
Address correspondence to Becky Knudson:
rebecca.a.knudson@odot.state.or.us

Department of Transportation report (*10*) provides analyses representing the "downside of doing nothing," but did not describe their methodology. Other states, such as Maine (*11*), outline critical activities that will not move forward under anticipated funding, but do not estimate the economic cost of these activities.

More common in the literature are examples of the benefits of funding specific, critical infrastructure maintenance and construction, such as a return-on-investment study looking nationwide (*12*). A recent report from Texas (*13*) calculates the return on investment of transportation funding and cites examples of specific projects. Cambridge Systematics produced a white paper for the NCHRP examining the economic benefits of transportation investments (*14*). Cost–benefit analyses for funding strategies are also found in the literature, such as a report by the Victoria Transport Policy (*15*). Studies following the approach presented in this paper were not found in the literature, but similar work may be occurring internal to agencies.

Methodology

The methodology for this study is based on planning-level forecast scenarios using three hypothetical budget increases in comparison with the current ODOT forecast budget for core highway asset spending categories: maintenance, pavement, bridge, enhance, seismic, and other investment. Budgeting for these categories is an integrated undertaking; investment tradeoffs are considered between the six categories, striking a balance between them to optimize returns on investment. Although the budget assumptions behind the scenarios include all six spending categories, the economic impact analysis prepared for this study only evaluates impacts associated with bridge and pavement expenditures, with the primary focus on bridges. Enhance projects typically address congestion, which requires detailed analysis, evaluating mode shift and land use. Current Oregon policies focus on preserving existing system conditions, which is the focus of this study. The four scenarios evaluated include:

Scenario 1: Current ODOT Forecast Budget for the state system.
Scenario 2: Expanded current investment required to keep Interstate 5 and Interstate 84 open to trucks.
Scenario 3: Hypothetical "What Would It Take" to preserve and repair the entire network of high-priority state highways, known as the Fix-It priority routes.
Scenario 4: Hypothetical "What Would it Take" to maintain current bridge and pavement conditions for the entire state-owned and operated system.

This analysis evaluates impacts of bridges restricted to weights below legal loads of 80,000 lb. The majority of trucks carrying freight on Oregon highways weigh 80,000 lb or more. Heavy trucks move over 70% of the commodities flowing within Oregon. Bridges restricted to lighter loads will require heavy trucks to detour or break up cargo into multiple trips or trucks, increase transportation costs, and impact businesses' access to markets. Depending on the number of bridges restricted and alternative routes available, the impact on accessibility varies by region and industry. Industries relying on heavy goods are most affected.

This study includes several features and assumptions, with a select few highlighted in this paper. For a more complete list, please see the full report available online (*16*):

- The modeled highway network is used to evaluate potential economic impacts of weight restricting bridges on the state-owned highway system. The modeled highway network does not include all highways and streets; most state highways are included along with a subset of local highways.
- The scenarios used for this study are based on priorities identified through the ODOT Statewide Transportation Improvement Program (STIP) Fix-It program (*17*), and the Seismic Plus Report (*18*), with a focus on bridges requiring work within the next 20 years.
- Bridge conditions were forecast using the ODOT Bridge Model, developed using Oregon bridge inspection and load rating data. Over 30 years of bridge inspection and load rating data were used as the basis to estimate bridge aging trends within the ODOT Bridge Model.
- Pavement conditions were forecast using the ODOT Pavement Management System Deterioration Model.
- The Oregon Statewide Integrated Model (SWIM) (*19*) was used to estimate impacts on the economy by incorporating forecast bridge restrictions onto the highway network. SWIM is an integrated economic, land-use, transportation forecast model, simulating the activity and market exchanges made by people and businesses.
- Inflation will impact the purchasing power of highway dollars over the next 20 years. By the year 2036 Oregon's 30 cent/gal state fuel tax purchasing power is reduced to an effective rate of 17.5 cents/gal. The current ODOT unadjusted "nominal" budget is US$10.9 billion over the 20-year period. Inflation reduces purchasing power of the 20-year budget by 20%, providing US$8.7 billion in 2016 adjusted "real" dollars.
- Impacts of Enhance projects, defined as investments improving mobility of people and goods by expanding transportation system capacity, are not included or represented in this modeling analysis other than for budget development.
- Scenario budgets were developed in a manner representing trade-off decisions made amongst the core highway asset categories of Maintenance, Pavement,

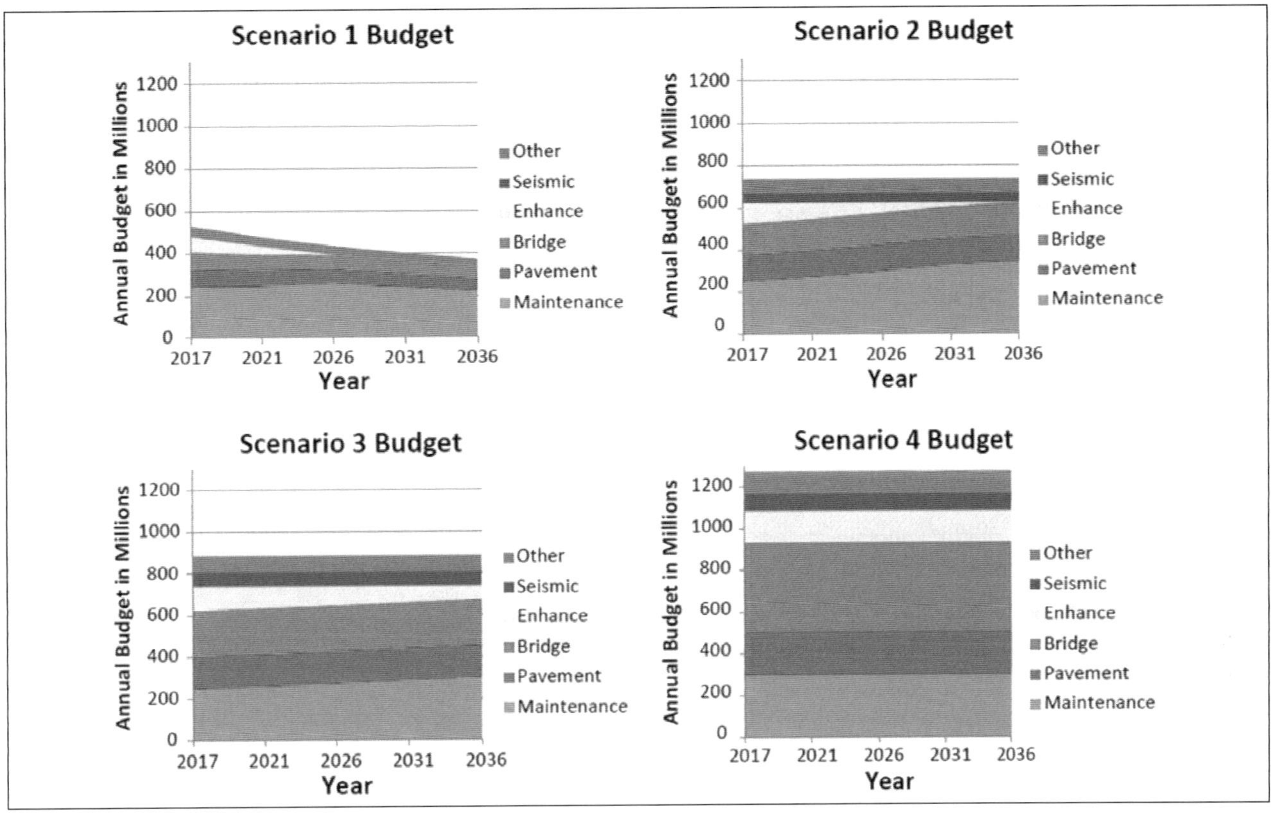

Figure 1. Scenario budgets.

Bridge, Enhance, Seismic, and Other investment. Figure 1 illustrates the four scenario budgets side-by-side for direct comparison. The details behind each budget are provided within the individual scenario descriptions.

Scenario Descriptions

For this study, the four scenarios were developed to explore tradeoffs associated with different levels of investment/disinvestment in the system. For these hypothetical scenarios, a simplified prioritization approach to corridor highway investment was used to evaluate high-level policy options and investment tradeoffs. Bridges were identified for work based first on route prioritization and then on the remaining service life reported out of the ODOT Bridge Model. For this analysis, culvert or landslide work is not included.

Scenario 1. Scenario 1 represents the expected 20-year forecast budget for highway spending for the core highway asset categories of Maintenance, Pavement, Bridge, Enhance, Seismic, and Other. As assets age, they require more preservation investment. When preservation investment does not occur, maintenance costs begin to increase. The scenario budget reflects this by shifting funding out of the Enhance budget and into the Maintenance budget. The annual Maintenance budget in this scenario increases by 3% each year by moving funds from the Enhance budget until the Enhance budget reaches zero by 2026.

By the year 2036 the 30 cents/gal state fuel tax purchasing power is reduced to an effective rate of 17.5 cents/gal due to inflation. Scenarios 2, 3, and 4 assume purchasing power is maintained over the 20-year period by an unidentified hypothetical inflation adjustment mechanism.

Scenario 1 investments focus on the I-5 corridor from Portland to Eugene, I-205, US26/US97 from the Portland metro region to the California border, OR58, and US20 from Bend to Ontario. These corridors connect most urban areas and provide east–west and north–south connectivity. Over the 20-year period the average annual nominal budget is US$545 million. After accounting for inflation, purchasing power declines to an average annual real budget of US$436 million. Thus, inflation is expected to reduce the purchasing power of the 20-year program budget of US$10.9 billion to US$8.7 billion, about a 20% reduction. Scenario 1 investment stipulates:

- 20-year bridge budget of US$1.36 billion preserves or replaces 130 bridges on the Fix-It routes, weight limits on 370 bridges to vehicles below 80,000 lb, with 98

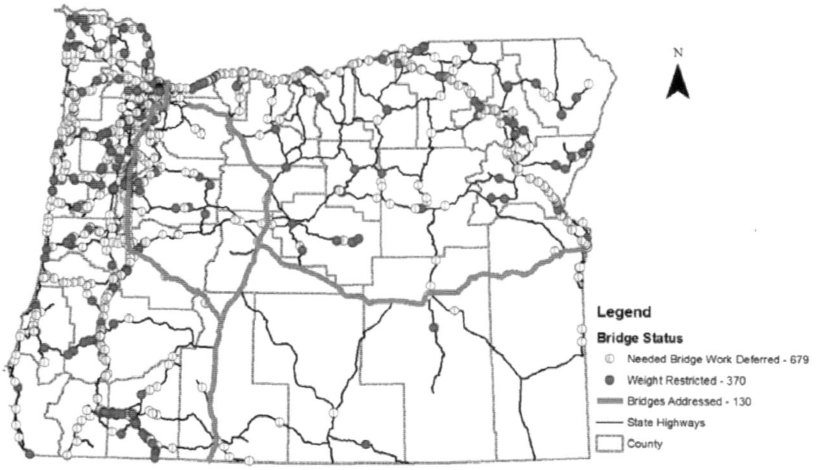

Figure 2. Scenario 1 bridge investments/disinvestments.

on the Fix-It priority corridors and 272 on other state highways; work is deferred for 679 bridges, including 254 bridges on Fix-It routes and an additional 425 bridges on other state highways, representing a US$7.3 billion backlog of bridge projects by the end of the 20-year period. Bridge investment is illustrated in Figure 2.

- 20-year pavement budget of US$1.36 billion, a US$2.6 billion short fall of the US$4 billion needed; by the end of the 20-year period the state would have to spend an additional US$4 billion to US$5.5 billion to return pavement conditions to current levels, requiring the state to pay more than double what it would have cost had investment occurred according to pavement management best practices.
- Overall, the backlog for state bridge and pavement projects over the 20-year period is anticipated to be greater than US$12 billion. This estimated backlog is likely less than actual need statewide, because it only includes state-owned facilities, does not account for repair costs of bridges not maintained, and does not include costs associated with culverts, landslides, and extreme events.
- For this scenario the Seismic budget is US$0. However, 39 bridges in this scenario are made seismically resilient in addition to the rehabilitation and replacement work completed. Although seismic projects are not explicitly prioritized in Scenario 1, work conducted on bridges for other structural reasons results in US97 and OR58 to be completed from the seismic preparedness perspective.

Scenario 2. Scenario 2 represents a larger investment budget designed to keep I-5 and I-84 open to trucks. It represents a 35% increase in the budget for core highway assets. Such an increase would require a 14 cent increase in the 30 cent state fuel tax and an equivalent increase in the heavy truck weight-mile tax. To reflect the disinvestment tradeoffs needed to address aging infrastructure, the Maintenance budget is increased 2%/year by shifting budget out of Enhance, until the Enhance budget reaches zero by the year 2034. Over the 20-year period the average annual budget is US$735 million expressed as 2016 dollars, making the 20-year program budget US$14.7 billion. Scenario 2 investment stipulates:

- 20-year bridge budget of US$3 billion preserves or replaces 355 bridges on the Fix-It routes, weight limits on 317 bridges to vehicles below 80,000 lb, with 45 bridges on the Fix-It priority corridors and 272 on other state highways; work is deferred for 507 bridges, including 82 bridges on Fix-It routes and an additional 425 bridges on other state highways, representing a US$5.7 billion backlog of bridge projects by the end of the 20-year period. Bridge investment is illustrated in Figure 3.
- 20-year pavement budget of US$2.54 billion, creating a US$1.5 billion dollar deficit over the 20-year period; by the end of the 20-year period the state would have to spend an additional US$2 billion to US$3.1 billion to return pavement conditions to current levels, requiring the state to pay up to double what it would have cost had investment occurred according to pavement management best practices.
- Estimated backlog of US$8.2 billion for bridge and pavement projects over the 20-year period.
- The 20-year Seismic bridge budget of US$980 million, combined with bridge replacements funded under the Bridge program, completes seismic retrofits and reconstructions identified as phases 1 and 2 of the Seismic Plus Program.

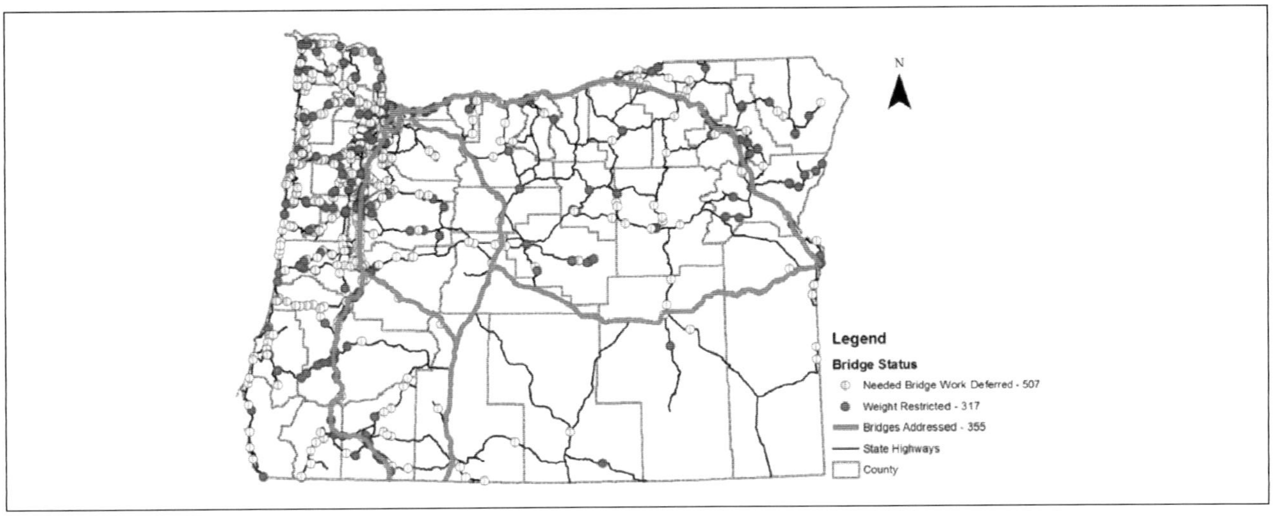

Figure 3. Scenario 2 bridge investments/disinvestments.

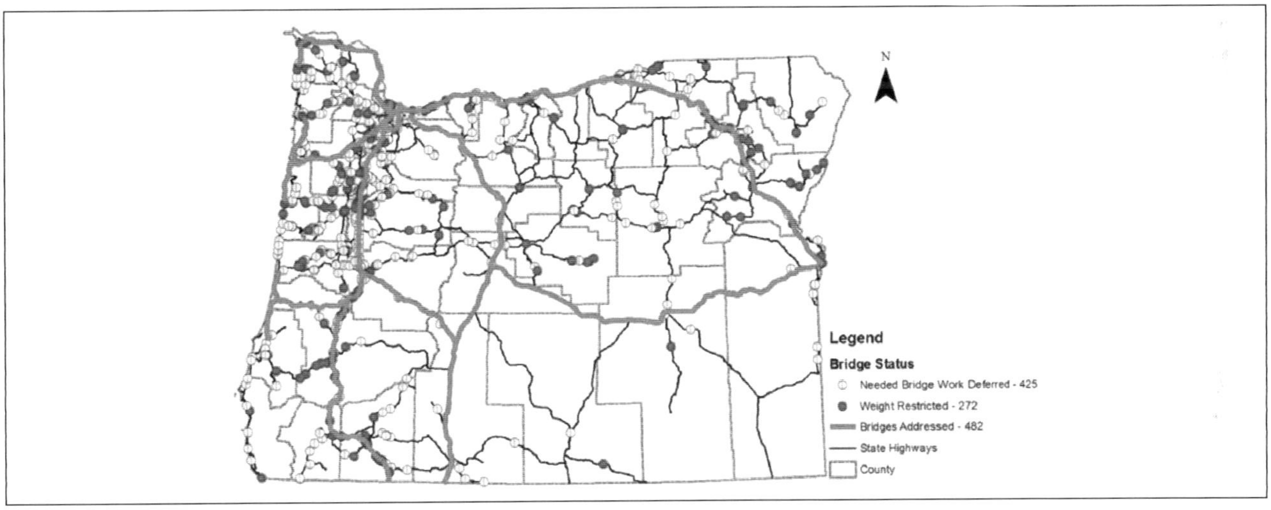

Figure 4. Scenario 3 bridge investments/disinvestments.

Scenario 3. Scenario 3 illustrates a hypothetical "What Would It Take" to preserve and repair the entire state Fix-It system. Such an increase would require a 24-cent increase in the state fuel tax and equivalent weight-mile tax. To reflect the disinvestment tradeoffs needed to address aging infrastructure, the Maintenance budget is increased 1%/year by shifting budget out of Enhance. Over the 20-year period the average annual budget is US$887 million expressed as 2016 dollars, making the 20-year program budget US$17.7 billion. Scenario 3 investment stipulates:

- A 20-year bridge budget of US$4.4 billion preserves or replaces 482 bridges on the Fix-It priority corridors, completing all of the Fix-It priority corridor bridge work; weight limits on 272 bridges on other state highways to vehicles below 80,000 lb; work is deferred for 425 bridges on other state highways, representing a US$4.3 billion backlog of bridge projects by the end of the 20-year period. Bridge investment is illustrated in (Figure 4).
- A 20-year pavement budget of US$154 million, creating a US$0.9 billion deficit over the 20-year period; by the end of the 20-year period the state would have to spend an additional US$1.3 billion to US$2 billion to return pavement condition to current levels, requiring the state to pay up to double what it would have cost if investment occurred according to pavement management best practices.
- Estimated backlog of US$6 billion for bridge and pavement projects over the 20-year period.

Figure 5. Scenario 4 bridge investments.

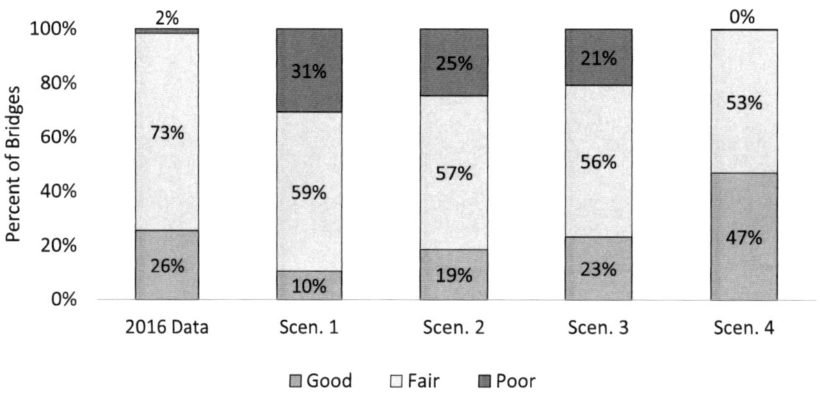

Figure 6. Bridge conditions: current and 2036 forecast by scenario.

- The assumed 20-year Seismic bridge budget of US$1.4 billion completes seismic retrofits and reconstructions identified for Phase 3, in addition to Phases 1 and 2 of the Seismic Plus Program.

Scenario 4. Scenario 4 illustrates a hypothetical "What Would It Take" to maintain current bridge and pavement conditions for the entire state system. Such an increase would require a 52-cent increase in the state fuel tax or equivalent weight-mile tax. In this scenario, all budget categories are stable and funded at levels needed to sustain the system efficiently and avoid weight restrictions. Scenario 4 investment stipulates:

- 20-year bridge budget of US$8.7 billion completes all needed bridge work, including 482 bridges on Fix-It route priority corridors, and 697 bridges on other state highways as illustrated in Figure 5.
- 20-year pavement budget of US$4 billion meets the needed resources to manage pavement according to pavement management best practices.
- The 20-year Seismic bridge budget of US$1.8 billion completes seismic retrofits and reconstructions identified in phases 1, 2, 3, and 4 of the Seismic Plus Program and includes one bridge from Phase 5 of the program.

Findings

For each scenario the numbers of good, fair, and poor bridges were forecasted. Figure 6 presents a summary of bridge condition by scenario. The weight-restricted bridges were coded onto the SWIM network to estimate economic impacts. The current forecast budget for bridge preservation and maintenance is expected to result in over 30% of Oregon state bridges rated as being in poor condition. Additional spending

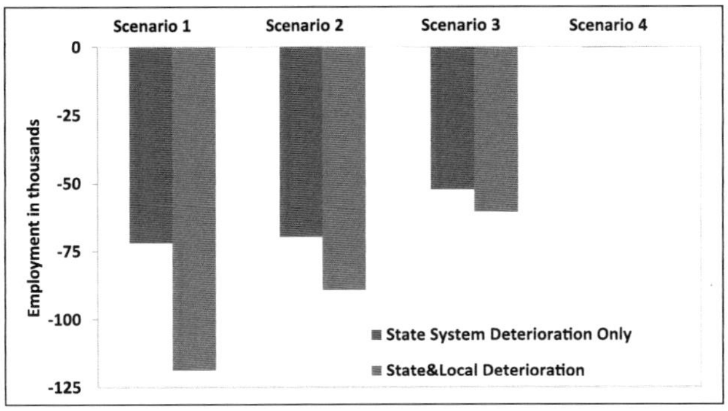

Figure 7. Forfeit employment statewide.

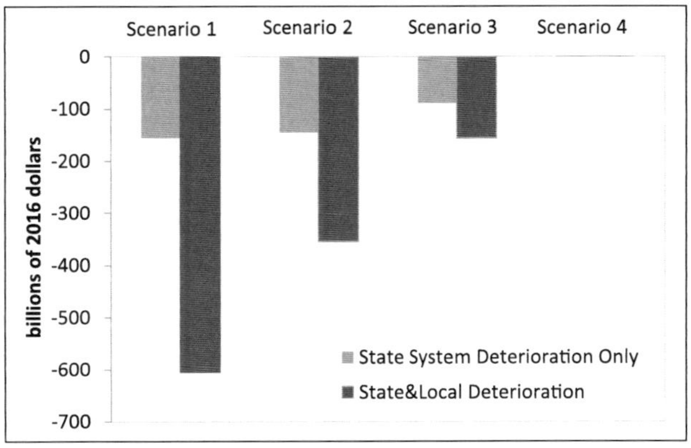

Figure 8. Cumulative losses in Oregon's gross domestic product.

represented in scenarios 2 and 3 increases the proportion of highly and poorly rated bridges due to the backlog of requirements.

A key finding of this study came about while evaluating traffic patterns from SWIM, revealing the importance of city- and county-owned facilities as alternative routes when state bridges are weight restricted. Currently, there is very little data available to represent local infrastructure inventory and conditions at the level of detail required for the tools used in this analysis. The analysis scenarios were run using several techniques to represent how non-state facilities may deteriorate as traffic is shifted from state routes to local routes. Thus, the results for impacts on employment and state production are presented with and without representation of local system deterioration to reveal a range of potential impacts.

Figure 7 illustrates the potential for job losses associated with rising transportation costs caused by bridge weight restrictions for the four scenarios, with employment levels of scenarios 1–3 compared with Scenario 4. Several patterns stand out:

- Failure to address the state system bridge needs could result in the forfeiture of 75,000 jobs, and up to 120,000 jobs if local infrastructure deterioration follows state patterns.
- Additional investment in the Fix-It priority corridors reduces potential job losses as system condition improves, owing to the reduction in transportation costs caused by weight-restricted bridges.
- The condition and operation of local transportation infrastructure matters to the Oregon economy. When state highway bridges are closed to heavy trucks, they reroute onto local roads in the model simulation. Consequently, when the modeled local transportation system does not account for deterioration, heavy trucks can use local detour routes without creating economic impacts. However, given that the age and condition of local bridges mirror the state highway system, this is not a realistic assumption. A rough estimate of the potential impacts of local conditions was characterized using a deterioration rate for the local system similar to the deterioration observed on the state highway system. Accounting for local system conditions and deterioration results in greater differences in employment forecasts.

Figure 8 illustrates the potential for cumulative losses in gross domestic product (GDP) over the 20-year period. The patterns here are similar to employment, with reduced production associated with system deterioration, and production higher in scenarios with additional investment. Accounting for local system deterioration increases impacts on cumulative production losses. The Scenario 1 reduction in total GDP over the 20-year period is over US$150 billion, which nearly quadruples when local system deterioration is accounted for. The Scenario 2 reduction is about 7% less than Scenario 1 when only state system deterioration is modeled. However, lost production is 60% lower in Scenario 2 when deterioration for state and local systems is included. The Scenario 3 level of investment reduces GDP losses by about 55% of Scenario 1 losses when accounting for state system deterioration only, but lost production is reduced by about 75% in Scenario 3 when including the impacts of local system deterioration.

These results demonstrate the importance of local government transportation infrastructure to the Oregon economy. About 15% of Oregon lane miles are owned by the state and carry about 60% of vehicle-miles-traveled (VMT) and 80% of truck traffic. The remaining 85% of lane miles are owned and maintained by city and county jurisdictions, carrying about 40% of VMT. Oregon roads and highways together, regardless of ownership, operate as a system with redundancy and alternative routes available in the short run. However, major closures and long-term heavy-vehicle detours would impact conditions on alternative routes, especially on facilities not designed for heavy trucks, or as a result of the increased use a detour may generate.

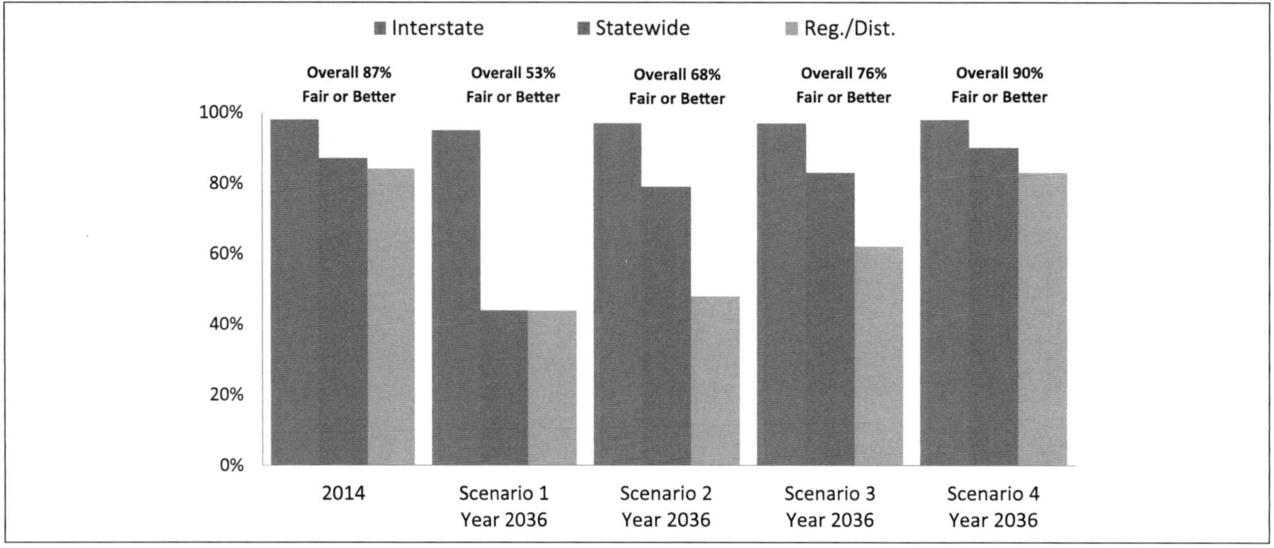

Figure 9. Pavement condition: 2014 and in the future by scenario and facility type.

Although the estimated economic impacts in this study focus on the effects of bridge weight restrictions, pavement investment is just as important for developing system redundancy. The goal of the ODOT pavement preservation program is to keep state highways in the best condition possible by following a life-cycle cost approach for preservation and maintenance. The typical cost of preserving pavement through resurfacing is about US$200,000 for each lane mile, whereas reconstructing that same lane mile after it fails can amount to as much as US$1.5 million (*20*).

An estimated US$200 million per year in current dollars for the next 20 years is sufficient to repair and maintain the condition of state highways in a "fair or better" condition. The current ODOT pavement budget of US$85 million per year is far less than what is needed.

Figure 9 illustrates current and forecast pavement conditions under all four scenarios. Preservation investments focus on the interstate system because of the high volume of vehicles and freight traveling at high speeds. Given the current 20-year funding forecast, by 2036 many segments will have reached the end of their structural life and require major rehabilitation or reconstruction to carry the projected traffic. The condition of the statewide system for primary routes and regional- and district-level routes will decline considerably.

Pavement condition impacts on vehicle operating costs were evaluated outside the modeling for this study. Preserving pavement condition at current levels would avoid higher user costs for private automobiles, commercial vehicles and heavy trucks. Vehicle operating costs rise as the pavement surface becomes rougher: fuel efficiency declines, tire wear and repair costs rise. Oregon has over 3.4 million registered light vehicles. Statewide VMT in 2015 was 35.9 billion, which includes Oregon-based and non-Oregon-based vehicles. Over 90% of the miles driven in Oregon are by light vehicles, 7% of the miles are by heavy trucks.

According to the 2015 Consumer Expenditures Survey (*21*) households spent on average 17% of their after-tax income on transportation. About one-third of transportation expenditures are for vehicle operating costs associated with fuel, oil, tires, maintenance, and repairs. As pavement condition deteriorates, more and more miles driven will be on rough pavement, causing vehicle operating costs to rise and household budgets to be impacted.

NCHRP Report 720 (*22*) conducted extensive research to estimate the impacts of pavement condition on vehicle operating costs for five vehicle categories: medium car, van, SUV, light truck, and articulated truck. For light vehicles, the report estimates vehicle operating costs rise by 23% when driving on rough pavement compared with smooth pavement. For trucks, the report estimates costs rise by 15%.

Table 1 presents the cost impact when 20% and 30% of statewide VMT occurs on rough pavement using the cost factors from NCHRP Report 720 and Oregon registered vehicle fleet characteristics for three Oregon categories: light vehicles, medium-heavy trucks, and heavy trucks. Total annual statewide vehicle operating costs would rise by about 4% if 20% of VMT was driven on rough pavement, which would increase statewide vehicle operating expenditures by US$300 million a year. Total statewide vehicle operating costs would rise about by 6% if 30% of VMT was driven on rough pavement, which would increase annual vehicle operating expenditures by US$450 million.

For businesses, higher transportation costs shift money away from more productive spending in areas related to investment and market expansion, and profit margins shrink. Rising costs such as this have the potential to reduce demand in key economic sectors in Oregon associated with business

Table 1. Potential Impact of Rough Pavement on Oregon Registered Vehicle Operating Costs

Potential impact of rough pavement on Oregon registered vehicle operating costs, (millions of 2015 dollars)					
Total estimated expenditures for year 2015 driving on smooth pavement		Total expenditures if 20% of miles driven on rough pavement	Total expenditures if 30% of miles driven on rough pavement	Change in cost with 20% rough	Change in cost with 30% rough
Light vehicles	US$5,309	US$5,558	US$5,682	5%	7%
Medium heavy	US$286	US$295	US$299	3%	5%
Heavy truck	US$1,494	US$1,539	US$1,561	3%	4%
Total statewide	US$7,088	US$7,391	US$7,542	4.3%	6.4%

Table 2. Scenario Summaries

	Scenario 1 Current funding	Scenario 2 Limited highway corridors	Scenario 3 Preserve priority corridors	Scenario 4 Meet needs
Economic impacts				
Forfeit jobs	75,000–120,000	70,000–90,000	50,000 –60,000	0
Lost GDP, US$ billions	$155–$605	$144–$355	$88–$156	0
State infrastructure condition				
Bridges repaired	130	355	482	1,179
Pavement % fair or better	53%	68%	76%	90%
Seismic budget, US$ billions	$0.0	$1.0	$1.4	$1.8
Seismic program	US97 & OR59	phases 1 and 2	phases 1–3	phases 1–4
Enhance budget, US$ billions	$0.4	$0.9	$1.8	$3.0
Other programs, US$ billions	$0.5	$1.3	$1.5	$2.0
20-year budget*, US$ billions	$8.7	$14.7	$17.7	$25.5
Bridge backlog, US$ billions (disinvestment)	$7.4	$5.7	$4.3	$0.0
Pavement backlog, US$ billions (disinvestment)	$4.6	$2.5	$1.7	$0.0
Total cost, US$ billions (budget + backlog)	$20.7	$22.9	$23.7	$25.5

*See scenario descriptions for detailed list of assumptions. Increased budgets are for 20-year core highway budget categories only: Maintenance, Pavement, Bridge, Seismic, Enhance, and Other, not the entire ODOT budget. Dollars are expressed in 2016 values.

investment for expansion, research, and development, and in consumer sectors. Households shift spending out of "quality of life" purchases, consumer goods, and savings to accommodate the rise in vehicle operating costs.

Conclusions

Under the current funding stream, the cost of maintaining and preserving existing facilities prevents investment in new and growing areas of the state because funding is insufficient to meet basic preservation and maintenance demands. Oregon must devise new ways to develop the efficient use and management of facilities and proactively prepare for disinvestment. Table 2 summarizes the costs and benefits evaluated using the four analysis scenarios developed for this study, illustrating trade-off decisions associated with different funding levels.

Though the transportation investment needs are large in scale, the additional investment cost is offset by the reduction in costs in the long run. Furthermore, the additional investment in the highway system will contribute to Oregon's economic competitiveness. The return on investment in the transportation system includes new jobs, more state output, lower transportation costs, and greater safety.

Oregon must continue to address transportation needs to accommodate future growth while maintaining economic competitiveness. The 2017 Oregon Legislature formed the Joint Committee on Transportation Preservation and Modernization to develop a transportation funding package. The Committee formed four modal workgroups to delve deeper into specific topic areas and report back to the full Committee. The Modal Workgroup on Maintenance, Preservation and Seismic used this study as a basis for developing their recommendation. In July of 2017 the Oregon Legislature approved a 10-year transportation funding package, the first comprehensive transportation funding bill to be passed in Oregon since 2009. This legislation serves as an example of how information developed using transportation modeling and analysis tools and methods can contribute to real-world decision making.

Acknowledgments

The authors would like to acknowledge the valuable insight and guidance provided by the Technical Advisory Committee for this project. We also wish to thank the ODOT team of technical experts, advisors, and consultant staff contributing to this study. This study was made possible as a result of the efforts of a collaborative team of talented and knowledgeable professionals.

References

1. Oregon Department of Transportation (ODOT). Oregon Transportation Plan. ODOT, Salem, Ore., September 20, 2006. http://www.oregon.gov/ODOT/Planning/Pages/Plans.aspx. Accessed September, 2006
2. Oregon Department of Transportation (ODOT). Oregon Freight Plan Modeling Analysis Technical Memo. ODOT, Salem, Ore., August, 2010. http://www.oregon.gov/ODOT/Planning/Documents/OFP-Modeling-Analysis.pdf. Accessed August, 2010.
3. Oregon Department of Transportation (ODOT). Estimated Economic Impact Analysis Due to Failure of the Transportation Infrastructure in the Event of a 9.0 Cascadia Subduction Zone Earthquake. ODOT, Salem, Ore., January 11, 2013. http://www.oregon.gov/ODOT/Planning/Documents/Cascadia-Subduction-Zone-Earthquake-Economic-Impact.pdf. Accessed January, 2013.
4. Oregon Department of Transportation (ODOT). Rough Roads Ahead: The Cost of Poor Highway Conditions to Oregon's Economy. ODOT, Salem, Ore., November 2014. http://www.oregon.gov/ODOT/Planning/Documents/Rough-Roads-Ahead-1.pdf. Accessed November, 2014.
5. Oregon Employment Department. Current Labor Force Statistics for Oregon. Presentation to Senate Interim Committee on Workforce. Oregon Employment Department, Salem, Ore., September 22, 2016. https://olis.leg.state.or.us/liz/2015I1/Downloads/CommitteeMeetingDocument/92803 Accessed October, 2016.
6. Oregon Office of Economic Analysis. Oregon Exports, Trade and Growth. Oregon Office of Economic Analysis, Salem, Ore., April 8, 2015. https://oregoneconomicanalysis.com/2015/04/08/oregon-exports-trade-and-growth/. Accessed October, 2016.
7. Oregon Business Council and Portland Business Alliance. The Cost of Highway Limitations and Traffic Delay to Oregon's Economy. Executive Summary. Oregon Business Council and Portland Business Alliance, Portland, Ore., March 2007. http://cdn.portofportland.com/pdfs/Trade_Trans_Studies_CostHwy_ExecSmry.pdf. Accessed April, 2007.
8. Wisconsin Department of Transportation. Keep Wisconsin Moving: Smart Investments Measurable Results. Transportation Finance & Policy Commission. Wisconsin Department of Transportation, Madison, Wis., January, 2013. http://wisconsindot.gov/Documents/about-wisdot/who-we-are/comm-couns/keep-wi-moving-report.pdf. Accessed October, 2017.
9. Cambridge Systematics Inc., D'Artagnan Consulting, LLP, and Indiana University. Study of Indiana Transportation Infrastructure Funding Mechanisms. Indiana Department of Transportation, Indianapolis, Ind., October, 2015. https://www.in.gov/indot/files/INDOT_finance_final_report.pdf. Accessed October, 2017.
10. The Boston Consulting Group. Investments in Washington State Transportation: Driving Economic Growth. Boston Consulting Group, Seattle, Wash., February, 2015. http://waroundtable.com/pdf/resources/Washington_Roundtable_Transportation_Study_FINAL.pdf. Accessed October, 2017.
11. Maine Department of Transportation. Connecting Maine: Planning Our Transportation Future. Maine Department of Transportation, Augusta, Maine, July, 2010. http://www.maine.gov/mdot/publications/docs/plansreports/connecting-mainefulldocument.pdf. Accessed October, 2017.
12. American Society of Civil Engineers. Failure to Act: Closing the Infrastructure Investment Gap for America's Economic Future. American Society of Civil Engineers, Reston, Va., 2016. http://www.asce.org/failuretoact/. Accessed September, 2016.
13. Ellis, D., and B. Stockton. The Benefits of Transportation Investment in Texas. Texas Department of Transportation, Austin, Tex., February, 2017. https://static.tti.tamu.edu/tti.tamu.edu/documents/0-6806-TTI-BS2.pdf. Accessed October, 2017.
14. Cambridge Systematics. Assessing the Economic Benefit of Transportation Infrastructure Investment in a Mature Surface Transportation System. Cambridge Systematics, Bethesda, Md., December, 2012. http://onlinepubs.trb.org/onlinepubs/nchrp/docs/NCHRP20-24(80)_FR.pdf. Accessed October, 2017.
15. Victoria Transport Policy Institute. Transportation Cost and Benefit Analysis: Techniques, Estimates and Implications. 2nd ed. Victoria Transport Policy Institute, Victoria, VC, Canada. http://www.vtpi.org/tca/. Accessed October, 2017.
16. Oregon Department of Transportation (ODOT). Rough Roads Ahead 2: Economic Implications of Deteriorating Highway Conditions. ODOT, Salem, Ore., February 2017. http://www.oregon.gov/ODOT/Planning/Documents/Rough-Roads-Ahead-2.pdf . February 2017. Accessed February, 2017.
17. Oregon Department of Transportation (ODOT). About the (Oregon) STIP. http://www.oregon.gov/ODOT/STIP/Pages/About.aspx. Accessed July, 2017.
18. Oregon Department of Transportation ODOT. Oregon Highways Seismic Plus Report. ODOT, Salem, Ore., October, 2014. http://www.oregon.gov/ODOT/Bridge/Documents/Bridge_seismic/2014_Seismic_Plus_Report.pdf. Accessed November, 2016.
19. Donnelly, R. (ed.)., WSP, ECONorthwest, HBA Specto Inc., and Resource Systems Group Inc. SWIM Version 2.5 Model Development Report. April 29, 2017. http://www.oregon.gov/ODOT/Planning/Documents/Statewide-Integrated-Model-Vers2-5.pdf. Accessed June, 2017.
20. Oregon Department of Transportation (ODOT). 2016 Pavement Condition Report. ODOT, Salem, Ore., February 2017, p. 13. http://www.oregon.gov/ODOT/Construction/Documents/Pavement/2016_condition_report.pdf. Accessed October, 2017.
21. Bureau of Labor Statistics. Consumer Expenditure Survey. Bureau of Labor Statistics, Table 1203. http://www.bls.gov/cex/tables.htm. Accessed July, 2016.
22. *NCHRP Report 720: Estimating the Effects of Pavement Condition on Vehicle Operating Costs*. HRB National Research Council, Washington, D.C., 2012.

The Standing Committee on Transportation Planning Applications (ADB50) peer-reviewed this paper (18-00649).

Article

TRR
JOURNAL OF THE TRANSPORTATION RESEARCH BOARD

Transportation Research Record
2018, Vol. 2672(46) 11–20

DOI: 10.1177/0361198118757980
journals.sagepub.com/home/trr
SAGE

How Autonomous Driving May Affect the Value of Travel Time Savings for Commuting

Felix Steck[1], Viktoriya Kolarova[1,2], Francisco Bahamonde-Birke[1,3], Stefan Trommer[1], and Barbara Lenz[1,2]

Abstract
Autonomous driving is being discussed as a promising solution for transportation-related issues and might bring some improvement for users of the system. For instance, especially high mileage commuters might compensate for some of their time spent traveling as they will be able to undertake other activities while going to work. At the same time, there are still many uncertainties and little empirical data on the impact of autonomous driving on mode choices. This study addresses the impact of autonomous driving on value of travel time savings (VTTS) and mode choices for commuting trips using stated-choice experiments. Two use cases were addressed – a privately owned, and a shared autonomous vehicle – compared with other modes of transportation. The collected data were analyzed by performing a mixed logit model. The results show that mode-related factors such as time elements, especially in-vehicle time and cost, play a crucial role for mode choices that include autonomous vehicles. The study provides empirical evidence that autonomous driving may lead to a reduction in VTTS for commuting trips. It was found that driving autonomously in a privately owned vehicle might reduce the VTTS by 31% compared with driving manually, and is perceived similarly to in-vehicle time in public transportation. Furthermore, riding in a shared autonomous vehicle is perceived 10% less negatively than driving manually. The study provides important insights into VTTS by autonomous driving for commuting trips and could be a base for future research to build upon.

Digitalization trends and rapid technology development have increased automation in all areas of daily life. Road vehicles are also becoming more technologically advanced in terms of automation with a continuing trend toward fully autonomous vehicles (*1*). There are high expectations placed on the technology, such as decreasing the number of road fatalities, reducing congestion, providing individual motorized mobility solutions to people currently not allowed or not able to drive, and to enable users to engage in other activities while driving (*1–3*). Certain user groups might benefit more than others, mainly depending on regular time spent traveling. This is especially the case for people, such as commuters, who routinely make long trips by car, have a limited time budget and therefore mostly a high willingness-to-pay (WTP) for saving travel time.

Commuting trips make up only a third of all trips in Germany but play a crucial role in road traffic as they determine peak travel demand (*4*). In recent years, commuting trips in Germany remained unchanged in terms of trip length [57% are shorter than 10 km (6.2 miles)], but increased slightly in terms of trip duration (22% take 30–60 min which corresponds to 4% increase in the share of trips with this duration) suggesting that more commuters are stuck in congestion on the way to and from work (*5*). Heavy traffic conditions at peak hours suggest that extensive commuting is often felt to be an exhausting and tedious task. A recent study on the relationship between mode choice and commuting stress found that car drivers have the highest stress levels compared with users of other modes. Furthermore, time consumption was among the most important subjective stressors for commuters driving on a daily basis (*6*). Therefore, an important benefit of having the opportunity to ride autonomously for commuters might be that they can compensate for time consumption commuting by using the time in a more efficient or more pleasurable way (*7, 8*).

The range of activities that can be performed during a trip depends, however, on the degree of automation of the vehicle. Referring to the definition given by the Society of Automotive Engineers (*9*), only Level 4 (the system in charge, some driving use-cases) and Level 5 (the system in charge, all driving

[1]German Aerospace Center, Institute of Transport Research, Berlin, Germany
[2]Humboldt-Universität zu Berlin, Geography Department, Berlin, Germany
[3]Technische Universität Berlin, Berlin, Germany

Corresponding Author:
Address correspondence to Felix Steck: Felix.Steck@dlr.de

use-cases) achieve the degree of independence from driving tasks that allow drivers to completely dedicate their attention to alternative activities. Thus, this study deals only with those levels of automation.

High-level automation will also enable new mobility services such as vehicles on demand either as individual "autonomous carsharing" service (ACS), similar to today's carsharing and taxi services, or as "autonomous ride sharing" (ARS), when pooling different trips together similar to uberPOOL. ARS services are expected to exhibit lower costs per mile at somewhat higher waiting times compared with ACS (*10*). These services could complement traditional public transport (e.g., solving the first/last mile problem) or act as a substitute where it is deficient today (*11–13*). From a user perspective these services could allow true door-to-door trips for individuals who do not currently have access to a car (*14*).

In summary, it can be expected that the car – be it privately owned or a shared vehicle – will become more attractive and, at the same time, available to broader user groups. This would lead to rebound effects, resulting in more vehicles on the road, more congestion and/or more vehicle miles traveled (*15–17*). Predicting changes in travel behavior and the traffic situation today is hard to do, but is more and more relevant in light of uncertainties about the future of mobility against the background of urbanization, demographic trends, and environmental challenges.

The aim of this paper is to analyze how autonomous driving may change mode choices for commuting trips. For this purpose, two different concepts of autonomous driving are considered. The first use case is privately owned autonomous vehicles (AVs) able to drive autonomously but with the option of switching off the autopilot. The second use case is shared autonomous vehicles (SAVs), which combine (Uber-like) the concepts of taxi and carsharing, where people can use a vehicle on demand. The results of the study should provide empirical insights into future modal-choice preferences for commuting trips.

Literature Review

The concept of value of travel time savings (VTTS) plays a crucial role in theoretical and empirical literature in transportation research. In accordance with microeconomic theory, individuals are supposed to make transportation decisions under the assumption that the daily time budget is constrained. Therefore, people choose whether they spend their time on one activity compared with another, or how much are they willing to pay to save the time spent in one particular activity (*18*). The subjective VTTS can be defined, therefore, as the WTP to reduce the travel time (*19*). VTTS usually depends on trip purpose and trip length and differs between modes of transportation. Studies on VTTS estimated higher values for commuting trips than for leisure or shopping trips (*20, 21*). In addition, the VTTS for commuting by car are lower in some studies, but in others higher, than for public transportation, and car passengers tend to have lower VTTS compared with car drivers (*20–22*). Furthermore, various empirical studies found that the VTTS of business travelers and commuters is higher in congestion than in free-flowing traffic (*18, 21, 23*). This suggests that even lower levels of automation might provide benefits for car users, for instance by enabling automated stop-and-go functions in dense traffic. Furthermore, it can be assumed that autonomous driving might potentially reduce VTTS for commuting trips in terms of perceiving the travel time less negatively.

Lately, a significant body of literature has addressed the possible impact of AVs on travel behavior (*7, 24, 25*). However, given the lack of empirical studies, potential reductions in VTTS are usually considered on the basis of plausible assumptions.

Filling this research gap is, however, a difficult task, as AVs are not currently available, so there is no existing behavioral data. Alternatively, it is possible to rely on stated preferences, but, in this case, respondents' lack of experience could affect the reliability of the results. Thus, it is advisable to center the analysis on high-level features, while acknowledging the limitations of the technique. Therefore, while attempting a detailed analysis of groundbreaking mobility options might prove difficult, focusing the analysis on potential reductions in the VTTS appears a more plausible task for respondents. To our knowledge, only a few stated-choice studies addressing this topic have been conducted to date.

Yap et al. address the use case of SAVs as egress transport for first/last mile trips in multimodal train trips considering time and costs for the trip as well as sharing levels (carsharing and ridesharing) (*11*). The results of the study suggest that first/last mile AVs can be attractive, especially for first-class train travelers. Furthermore, the sensitivity of users to in-vehicle time is higher in autonomous compared with manually driven vehicles, resulting in higher VTTS, which the authors attribute to attitudinal and perceptual concerns about the technology. Along these lines, the results by Winter et al. show strong differences between early and late adopters (with a clear preference for SAVs in the early adopters group) in the context of modal choice, while including an SAV alternative (*26*). The results of Krueger et al. on the adoption of SAVs show a similar trend (*27*). The authors found that service attributes including travel time, waiting time, and fares, had a strong impact on mode choices, and individual-specific characteristics, such as age and an individual's modality style also had a significant effect.

All three studies, while providing initial empirical insights into user preferences regarding AVs, focus on the introduction of SAVs as an alternative to current modes of transportation. In doing so, they ignore nonmotorized alternatives, such as walking or cycling, but also the option of privately owned AVs. Therefore, it is not possible to gain from these

Table 1. Attributes and Attributes' Levels

Transport mode	Attribute	Levels
Walk	Time	−30%|−10%|+20% reference Time [Speed: 4.9 km/h]
Bike	Time	−30%|−10%|+20% reference Time [Speed: 15 km/h]
	Access time	2 min | 5 min
Autonomous vehicle (AV)	Time	−30%|−10%|+20% reference Time [Speed: between 26 km/h and 68 km/h, distance dependent]
	Waiting time	2 min | 5 min | 10 min
	Cost	−30%|−10%|+20% current costs [€0.20/km]
Shared autonomous vehicle (SAV)	Time	−30%|−10%|+20% reference Time [Speed: between 26 km/h and 68 km/h, distance dependent]
	Waiting time	2 min | 5 min | 10 min
	Other passengers	alone/other passengers
	Cost	−30%|−10%|+20% reference costs "alone" [€0.20/km] -30%|−10%|+20% reference costs "other passengers" [€0.20/km]
Public transportation	Time	−30%|−10%|+20% reference Time [Speed: between 18 km/h and 51 km/h, distance dependent]
	Access time	2 min | 5 min | 10 min
	Waiting time	2 min | 5 min | 10 min
	Cost	−30%|−10%|+20% current costs [between €1.5 and €6, distance dependent]

Note: 1 kilometer = 0.62 miles; 1 euro = US$1.10 (exchange rate during study period: May 5, 2017).

studies' insights on the willingness to use privately owned AVs compared with SAVs.

Some recent studies also address user preferences toward privately owned AVs. However, these studies focus on the impact of autonomous driving on car ownership or possession of a public transportation pass. Becker and Axhausen used a stated choices approach to assess the impact of SAVs and privately owned AVs on mode choices (*28*). Their pilot study with 62 participants suggests a decrease in car ownership rate by introducing autonomous driving, especially as a sharing service. Another stated-choice study on the impact of AVs for commuting trips found that, besides cost, various attitudinal variables, such as technology interest and enjoyment of driving, influence the user preferences toward the technology (*29*). However, the study focuses more on long-term choice decisions than on influencing factors on trip mode choices.

In summary, no studies were identified that focused on the evolution of VTTS related to the introduction of AVs, nor studies that addressed both privately owned AVs and SAVs simultaneously, and compared with all other relevant modes of transportation.

Methods

Study Design

To address the research questions, an online survey was conducted using a questionnaire with the following structure: questions on existing mobility behavior, questions on the commuting trip the person usually takes (i.e., reference trip), a short introduction to the concept of autonomous driving, a discrete-choice experiment (DCE), questions on willingness to purchase and pay for AVs, as well as sociodemographics.

The study design was based on an earlier methodological approach that combines revealed and stated preference data (*30, 31*). In the revealed preference part of the questionnaire, the respondents were asked to describe a recent trip. In the stated-choice experiments, hypothetical mode-choice situations for the same trip were constructed using the individual trip length of the respondents. In each choice situation, the time and the cost for the trip were reduced or increased around reference values using estimated average speeds and costs for each mode of transportation. The choice experiment consisted of eight choice situations in which the respondents had to choose between one of the following five transportation options: walk, bike, public transportation, privately owned AV, and an SAV. The SAV was called "driverless taxi" to provide a better understanding of the concept to the participants. The attributes and their levels used in the experiments are summarized in Table 1. It was assumed that an AV drives up to users, drops them off and finds a parking spot by itself. Therefore, access and egress time for the AVs were excluded as attributes and waiting time was considered.

To present realistic alternatives to the study participants, "average speeds" and "cost per transportation mode" were used for the German case. Average speeds were estimated using the German national household travel survey from 2008, called MiD 2008 (*4*). The costs per kilometer for the private car were drawn from the Allgemeiner Deutscher Automobil-Club (*32*). Only fuel and maintenance cost were taken into account. Costs related to the purchase of the vehicle or parking cost were not considered. The kilometer price for the SAVs followed existing analysis (*10*). The cost for public transportation was drawn from existing rates for public transportation systems in Germany. Fixed costs were used for different distance classes with a minimal price of €1.50. Season, annual or student tickets for public transportation were not considered.

To enhance the data quality of the experiments by maximizing the information obtained from each choice situation, a Bayesian efficient design was created, using the software Ngene (*33*). An efficient design is recommended when some initial information about the value of the parameters is available prior to the field test, as it can improve the design significantly and reduce the standard error (*34*). In the current study, the prior values for the estimation of the efficient design were drawn from a pretest with 30 respondents. The design was optimized for short, and medium, and long trips so as to consider the effect of trip distance on trip and mode-related attributes.

Introduction of the Concept of Autonomous Driving

The two concepts of autonomous driving privately owned and shared AVs were presented to the study participants in two short animated videos before the choice experiment. In the first video the main character, Ms. Schmidt, calls her vehicle using an app on her phone, rides to her preprogrammed destination, gets out of the car on arrival, and the vehicle continues to drive autonomously and parks itself. During the trip she can decide whether she prefers to drive manually or ride autonomously. In the second video, the concept of an SAV is introduced. It is shown that one can order the vehicle, ride autonomously to one's destination, get out of the car and the vehicle then drives on to collect its next passenger(s). The main difference between the two introduced concepts was that, in the privately owned vehicle, there was an option to switch off the autopilot. In the SAV there was no steering wheel or brakes, it could not be driven manually. The two concepts were presented as neutrally as possible (without using evaluative adjectives) so as to influence the preferences toward autonomous driving as little as possible.

To find out if respondents preferred to drive their hypothetical privately owned vehicles autonomously or manually, an additional question was added with a Likert scale related to this preference after the choice experiment. Based on the responses, two dichotomous variables were created that indicated whether they preferred to use their privately owned vehicles autonomously or use them manually.

Implementation and Sample

For the online implementation of the questionnaire including the choice experiment the software Sawtooth was used (*35*). Survey participants were recruited using a professional panel service. A sample of 485 respondents representative of Germany by age and gender was recruited. The sample included car users as well as noncar users and was limited to participants older than 18 years. The duration of filling in the online survey was 13 min on average. The respondents were randomly selected to provide information about one of three different trip types – commuting trips, shopping trips and leisure trips. However, in this paper a reduced sample size of 172 respondents was used as the rest of the sample reported trips other than commuting.

A comparison between the reported commuting trips of our sample and commuting trips from MiD 2008 (*4*) shows that the key parameters are largely similar (see Table 2). A critical point is the overrepresented public transport use and by contrast, the underrepresented car use in our sample. The mode-specific distances and times of commuting trips fit quite well. However, by using trip length and trip duration as reference parameters of the presented choice experiments, the existing data seem to be suitable.

Analysis Method

The most common alternatives in mode choice with multiple alternatives are the multinomial logit (MNL) and the more advanced mixed logit (ML) models (*36*). The MNL model developed and described by McFadden estimates the probability of each individual n selecting alternative i (*37*). Here it is assumed that n assigns a given utility to every alternative i in the sampling, opting for the alternative that maximizes the expected utility. Assuming additive linearity, the expected utility is given by the following expression:

$$U_{n,i} = \beta_n X_{n,i} + \varepsilon_{n,i} \qquad (1)$$

Here $X_{n,i}$ is a vector of explanatory variables including the attributes of the alternatives as well as socioeconomic characteristics of the respondent, and β_i are parameters to be estimated. The error term $\varepsilon_{n,i}$ represents a stochastic component, accounting for all relevant attributes that are ignored by the modeler. An MNL imposes the condition that $\varepsilon_{n,i}$ follows an independent and identically distributed (iid) extreme value type 1 distribution (*37*). However (and because of the restriction imposed upon the distribution of the stochastic elements), the MNL does not allow for consideration of heterogeneity among respondents, nor for capturing the pseudo-panel nature of our data, that is, eight choice sets per person. Thus, an ML was employed to relax the assumptions that the coefficients are the same for all individuals (*38, 39*) and to allow correlation across choice situations (*36, 40*). The utility function of an ML with panel data can be extended as follows:

Table 2. Comparison of the Commuting Trips between the German National Travel Survey, MiD 2008 (4), and the Study Sample

	Walk	Bicycle	Car	Public transport	Mean
			German National Travel Survey		
Modal split (in %)	7	10	70	12	–
Commuting time (in min)	11	15	26	53	27
Commuting distance (in km)	0.9	3.5	20.0	25.8	17.7
			Study sample		
Modal split (in %)	9	8	60	23	–
Commuting time (in min)	17	14	24	46	27
Commuting distance (in km)	2.0	3.8	19.7	25.2	18.1

Note: 1 kilometer = 0.62 miles.

$$U_{n,i,t} = bX_{n,i,t} + \eta_n X_{n,i,t} + \varepsilon_{n,i} \quad (2)$$

Here, the coefficient vector β_i from Equation 1 is expressed as $b_i + \eta_n$. In this framework, b_i accounts for the population mean and η_n is a random term following a distribution to be established by the analysis with a given mean (normally 0) and density to be estimated. This allows for accounting for different valuations of $X_{n,i}$ across individuals. t represents the different choice situations a given individual n is confronted with, and therefore $b_i + \eta_n$ is not assumed to vary across different choice situations t, taking the pseudo-panel effect into account (i.e., the valuation of the attributes remain constant for all observations associated with the same individual). The ML probabilities of choosing given alternative i is, consequently, a weighted mean of the MNL probabilities at a specific η, weighted over the distribution of η.

$$P_{n,i} = \int L_{n,i}(b,\eta) f(\eta) d\eta \quad (3)$$

In Equation 3 the choice probability $L_{n,i}$ represents the MNL probabilities for a given value of η. Due to the fact that an individual is faced with t choice situations, the probability of observing a given sequence of choices is given by the following expression:

$$L_{n,i}(\Omega) = \prod_{t=1}^{T} \left(\frac{e^{\beta x_i}}{\sum_{j=1}^{J} e^{\beta x_j}} \right) \quad (4)$$

Model Specification

To obtain the final model specification, an iterative procedure was used. In the first step of the analysis, an MNL was estimated only considering time and cost parameters. Afterwards, socioeconomic variables were introduced (solely significant socioeconomic variables were ultimately included in the models). The final specification of the model considered the following explanatory variables:

TT_i: *travel time of mode i (in min)*
TC_i: *travel cost of mode i (in €)*
SR: *dummy for shared ride for driverless taxi*
MAN: *dummy for individual who prefers driving privately owned AV manually*
AUT: *dummy for individual who prefers driving privately owned AV autonomously*
LH: *dummy for license holder*
PT_CARD: *dummy for holder of a public transport pass*
AGE^{middle}: *dummy for middle-aged individual (between 30 and 50 years old)*
AT_i: *access and egress time for mode i (in min)*
WT_i: *waiting time for mode i (in min)*
INC: *dummy for income class [low: up to €1.500 (US$1.650), middle: €1.500–€3.000 (US$1.650–US$3.630), high: more than €3.000 (US$3.630)]*

All explanatory variables are assumed to have a linear additive impact on the utility functions, although not all of them affect the utility of all alternatives. Furthermore, it is assumed that the alternative-specific constants (ASC) and the valuation of the generalized travel time (see below) exhibit stochastic variations across individuals. The distribution of the β_i associated with the ASC and the generalized travel time is assumed to be normally distributed. The β parameter associated with the cost of the alternatives is assumed to exhibit variation among income classes.

To consider a decreasing marginal utility of time and cost on mode choices, a Box-Cox transformation was used (41). From a behavioral standpoint, this might – especially in the case of commuting trips – provide important insights on time perception and VTTS depending on travel distance. The considered transformations are depicted in Equation 5.

$$\beta_{\text{Time},i} \cdot \frac{(TT_i + \beta_{\text{Acc}} \cdot AT_i + \beta_{\text{Wait}} \cdot WT_i)^{\lambda_{\text{Time}}} - 1}{\lambda_{\text{Time}}}$$
$$\text{and } \beta_{\text{Cost}} \cdot \frac{(TC_i)^{\lambda_{\text{Cost}}} - 1}{\lambda_{\text{Cost}}} \quad (5)$$

Here, the expression considered in association with the time parameter represents the generalized travel time, which takes into account that access and waiting time are perceived differently from in-vehicle travel time. Here, β_{Acc} and β_{Wait} are also parameters to be estimated, which also exhibit variability across individuals. However, in contrast to β_{Time} , the distribution of β_{Acc} and β_{Wait} is considered to be uniform, so as to avoid problems with negative values inside the Box-Cox transformation.

Finally, two ML models were estimated, one of which did not consider nonlinearity, and the other considered the Box-Cox transformation. As previously mentioned, parameter variability across individuals was only considered for time-related variables (i.e., travel time, access and egress time, and waiting time) and the ASCs. The estimation of the models was preformed using PythonBiogeme (*42*). The distributions of the random parameters were simulated by using 5,000 modified Latin hybercube sampling draws (*43*).

Results

Estimated Model Coefficients

The results of the two final estimated ML models are summarized in Table 3. In general, the coefficients exhibit the expected signs and plausible values. A significantly better model fit was obtained by modeling possible nonlinearity for the time and cost parameters, χ^2 (2, N=172)=9.65, p<.01. Therefore, our results confirm the existence of decreasing marginal utilities.

Overall, the results show that cost and travel time elements influence mode choices significantly, both having an expected negative impact. The coefficients in Model 2 were higher than in Model 1 but had similar relations to each other, suggesting stable tendencies.

The generalized time coefficients show differences between the modes. Travel time in privately owned AVs was perceived less negatively by people using the automation function on commuting trips compared with people driving manually. The segmentation of the respondents into the two groups was based on two dichotomous variables, which specify if a person prefers to drive manually or ride autonomously. The across-population variability of the estimated coefficients suggests a wider heterogeneity between driving AVs automatically than manually. Moreover, riding autonomously to work was perceived less negatively than the travel time of any of the other available motorized alternatives. However, the differences were not statistically significant.

When considering the ASCs, the general preference for SAVs was significantly lower compared with privately owned vehicles; however the mode was more attractive than public transportation. At the same time, exploration of the travel time coefficients suggests that riding autonomously in an SAV was perceived less negatively than driving, but was less attractive than riding autonomously with a privately owned vehicle.

However, a comparison between the modes is only possible when considering all time elements, including waiting and access/egress time. The coefficients for these two time elements were estimated in relation to in-vehicle time. While there were no major differences between access and in-vehicle travel time (access time was perceived as slightly more negative), waiting time was perceived 2.12 to 3.28 times more negatively (depending on the model) than the in-vehicle time.

Furthermore, as expected, there was a relationship between cost sensitivity and household income. People on low incomes were more cost-sensitive, perceiving travel cost more negatively, than people with middle or high incomes. This is reflected in the WTP differences described in the following section.

The analysis of the perception of autonomous carsharing compared with autonomous ridesharing (represented by β_{SHARED}) did not provide any statistically significant evidence on whether people would prefer to share a ride with others or to ride alone in an SAV. This suggests the sharing aspect has a smaller role compared with other factors.

Regarding the impact of sociodemographic factors, only the variables found to exhibit a statistically significant effect were included in the final model. No significant effect of gender on mode preferences in the final estimations was found. On the subject of age, the analysis shows that middle-aged people (between 30 and 50 years old) were less inclined to walk or cycle to work than younger or older people. Possession of a public transportation pass positively influenced preferences for that mode. Furthermore, people who possess a driver's license were less inclined to walk to work. No socioeconomic variables that directly related to preferences toward AVs were found.

Estimation of VTTS

As previously mentioned, the main objective of the analysis is to establish the differences among the VTTS according to transportation mode when AVs are available. This would allow us to establish the extent to which relieving users from the task of driving might impact on perceptions of time.

Establishing the VTTS was straightforward for Model 1, as we considered constant marginal utilities of both travel time and costs, so that the VTTS could be established in accordance with the following expression:

$$\text{VTTS} = \frac{\partial U_i / \partial \text{TT}_i}{\partial U_i / \partial \text{TC}_i} = \frac{\beta_{\text{Time},i}}{\beta_{\text{Cost},n}} \tag{6}$$

For Model 2, considering decreasing marginal utilities for both travel time and cost, the VTTS depends on the actual travel time and cost experienced by the user, as in the following expression:

Table 3. Results of the Two Mixed Logit Model Estimation

	Model 1: Mixed logit		Model 2: Mixed logit with a Box-Cox transformation for time and cost	
Coefficient	Estimated value	*t*-value	Estimated value	*t*-value
ASC_{PED}	11.9	*(4.02)*	14.6	*(4.75)*
ASC_{BIKE}	4.42	*(4.62)*	8.25	*(4.32)*
ASC_{PT}	−3.39	*(−3.27)*	−2.68	*(−2.07)*
ASC_{SAV}	−1.74	*(−2.89)*	−1.62	*(−2.29)*
η_{-PED}	0.857	*(0.71)*	−0.372	*(−0.35)*
η_{-BIKE}	3.64	*(5.86)*	3.53	*(5.73)*
η_{-PT}	2.81	*(3.35)*	2.25	*(2.92)*
η_{-AV}	1.38	*(2.88)*	−0.559	*(−0.82)*
η_{-SAV}	−1.65	*(−4.01)*	−1.83	*(−5.08)*
β_{TIME_PED}	−0.423	*(−4.56)*	−1.31	*(−1.87)*
η_{-TIME_PED}	−0.168	*(−4.34)*	−0.292	*(−2.34)*
β_{TIME_BIKE}	−0.314	*(−7.51)*	−1.35	*(−1.90)*
η_{-TIME_BIKE}	−0.116	*(−5.51)*	−0.394	*(−2.45)*
β_{TIME_PT}	−0.0825	*(−3.60)*	−0.402	*(−1.62)*
η_{-TIME_PT}	0.0703	*(4.19)*	0.254	*(2.12)*
$\beta_{TIME\_AV\_AUTONOM}$	−0.0784	*(−3.69)*	−0.307	*(−1.65)*
$\eta_{-TIME\_AV\_AUTONOM}$	0.062	*(2.48)*	0.213	*(2.17)*
$\beta_{TIME\_AV\_MANUAL}$	−0.114	*(−5.84)*	−0.442	*(−1.93)*
$\eta_{-TIME\_AV\_MANUAL}$	−0.0355	*(−1.66)*	−0.109	*(−1.76)*
β_{TIME_SAV}	−0.102	*(−4.68)*	−0.403	*(−1.84)*
η_{-TIME_SAV}	0.0183	*(0.80)*	0.0324	*(0.51)*
$\beta_{WAIT\ (uniform\text{-}bottom)}$	1.08	*(3.83)*	1.01	*(4.05)*
$\eta_{-WAIT\ (uniform\text{-}top)}$	3.28	*(3.82)**	2.12	*(4.45)**
β_{ACC}	1.08	*(3.22)*	1.07	*(3.97)*
$\beta_{COST\_LOW\_INC}$	−1.14	*(−5.72)*	−1.52	*(−4.49)*
$\beta_{COST\_MID\_INC}$	−0.947	*(−6.1)*	−1.24	*(−3.54)*
$\beta_{COST\_HIGH\_INC}$	−0.543	*(−5.61)*	−0.79	*(−3.24)*
β_{SHARED}	0.0191	*(0.07)*	−0.033	*(−0.13)*
λ_{COST}	–	–	0.787	*(5.89)*
λ_{TIME}	–	–	0.566	*(3.50)*
β_{PT_CARD}	1.43	*(1.71)*	1.98	*(2.54)*
$\beta_{LICENCE\_PED}$	−4.74	*(−2.22)*	−4.6	*(−2.91)*
$\beta_{MID\_AGE\_PED}$	−4.14	*(−2.70)*	−4.11	*(−3.62)*
$\beta_{MID\_AGE\_BIKE}$	−3.27	*(−3.09)*	−3.62	*(−3.40)*
Model Fit				
Log-likelihood (final)	−948.011		−943.187	
Estimated parameters	32		34	
Observations	1,376		1,376	

Note: * The *t*-values refer to the bottom level of the uniform distribution.

$$\text{VTTS} = \frac{\partial U_i / \partial \text{TT}_i}{\partial U_i / \partial \text{TC}_i} = \frac{\beta_{\text{Time},i} \cdot (\text{TT}_i + \beta_{\text{Acc}} \cdot \text{AT}_i + \beta_{\text{Wait}} \cdot \text{WT}_i)^{\lambda_{\text{Time}}-1}}{\beta_{\text{Cost},n} \cdot (\text{TC}_i)^{\lambda_{\text{Cost}}-1}} \quad (7)$$

Therefore, it was only possible to calculate an average for the considered population. Furthermore, as the marginal utility of the price depends on the actual cost, the VTTS would exhibit slight variation (<5% in our case) depending on alternative used as reference. In this work, we have considered the marginal utility of the cost of SAVs as the reference to establish the VTTS. The estimated values are summarized in Table 4.

Table 4. Estimated VTTS for Different Modes of Transportation and Income Classes (in Euro/Hour)

	Walk	Bike	Public transport	AV autonomously	AV manually	SAV
Model 1: Mixed logit						
Low income	22.26	16.53	4.34	4.13	6.00	5.37
Middle income	26.80	19.89	5.23	4.97	7.22	6.46
High income	46.74	34.70	9.12	8.66	12.60	11.27
Model 2: Mixed logit with a Box-Cox transformation for time and cost						
Low income	8.88	13.41	3.93	3.74	5.39	4.85
Middle income	10.88	16.44	4.81	4.59	6.60	5.94
High income	17.08	25.88	7.56	7.20	10.36	9.32

Note: 1 euro = US$1.10 (exchange rate during study period: May 5, 2017).

The results for the VTTS reflect the results from the estimations presented above. People with a high income had a higher WTP for saving commuting travel time. Here, again, the VTTS for people who prefer autonomously driving privately owned AVs was lower than the VTTS of people driving manually by 31% in both models. It reflects the perceived benefits of relieving the user from driving tasks and allowing them to dedicate their attention to activities deemed more meaningful. The VTTS for driving autonomously was in the range of VTTS for in-vehicle time in public transportation, suggesting a similar perception for both modes of transportation. However, it did not include waiting and access/egress time, which can be, as estimated above, up to two or three times more negative than in-vehicle time (this phenomenon negatively affects the perception of public transportation). At the same time, the VTTS for SAVs was slightly higher than autonomously driven vehicles and public transportation, but still 10% lower than for driving a car by oneself. Therefore, using an SAV may be deemed more attractive than driving manually to work (although relying on SAVs may also involve waiting time).

Conclusions

The main aim of this study was to analyze how autonomous driving may affect the subjective VTTS for commuting trips. For this purpose, a DCE was conducted and the data were analyzed using an ML model.

First, the results provide empirical evidence supporting the assumption that autonomous driving will potentially reduce the VTTS for commuting trips, that is, it will be an attractive function for people making regular commuting trips. Moreover, the VTTS for two different possible uses of autonomous driving, namely privately owned AVs and SAVs, were estimated for different income classes and contrasted with alternative modes of transportation. Our results suggest that driving autonomously leads to a reduction of 31% in the VTTS compared with driving manually, and is perceived similarly to the VTTS of in-vehicle time in public transportation (waiting and access/egress time are perceived more negatively in public transportation).

Second, when considering the preferences toward SAVs, we found that travel time spent in SAVs is perceived 10% less negatively than driving manually. However, riding privately owned AVs seems to be more attractive than using SAVs. In general, the preference for using a privately owned vehicle in the sample seems to be higher than using shared vehicles; at least for regular commuting trips. Even though the VTTS in SAVs seem to be a little higher than the in-vehicle VTTS for public transportation, the calculation does not include the longer waiting and access/egress times associated with public transportation (and the fact that the travel time in public transportation is usually greater than by car). This suggests there is potential for SAVs to be an alternative (or complementary service) for public transportation.

Regarding different user perceptions toward autonomous ridesharing compared with autonomous carsharing, our study does not offer conclusive results. However, we suggest that users' concerns about sharing a ride with strangers are possible. Thus, attitudes toward sharing a ride have to be considered in future work, for instance using a sample of people with ridesharing experience, such as users of uberPOOL, or of privately organized ridesharing.

The main limitation of the study is related to possible hypothetical bias as AVs are not currently available. Therefore, providing realistic answers may be difficult for the respondents, as they do not have direct experience with the technology. Therefore, while acknowledging the limitations of the technique, we centered the analysis on a high-level feature, the VTTS, which may have been easier for the respondents to internalize.

In all, the study provides important empirical evidence and insights into how autonomous driving might affect mode choices and valuation of travel time for commuting trips. This study thus lays the groundwork for the possible impacts of introducing AVs on the valuation of travel time, which future research can build upon. Along the same lines, the study provides empirical evidence supporting the reduction of the VTTS, considered by many authors in simulation exercises.

Future research should focus on other relevant determinants of mode choice, and also on understanding the perception of in-vehicle time for autonomous driving, which has not been covered in this study. For instance, geographic variables can play an important role in mode-choice decisions and should be explored in future research. In any case, caution is required, as respondents may be overwhelmed when confronted with groundbreaking technologies they are not familiar with. Thus, the analyst should focus their efforts on aspects the respondents can deal with. Another avenue for future research may be in understanding the determinants behind user preferences and perceptions. Therefore, further work on users' attitudes and needs as well as perceived individual benefits of automation might be crucial in understanding commuters' decision-making processes.

References

1. Fagnant, D. J., and K. Kockelman. Preparing a Nation for Autonomous Vehicles: Opportunities, Barriers and Policy Recommendations. *Transportation Research Part A* Vol. 77, 2015, pp. 167–181.
2. Anderson, J. M., N. Kalra, K. D. Stanley, P. Sorensen, C. Samaras, and O. Oluwatola. *Autonomous Vehicle Technology – A Guide for Policymakers*. RAND Corporation, Santa Monica, CA, 2014.
3. Litman, T. Ready or Waiting? *Traffic Technology International*, 2014, pp. 35–42.
4. DLR, and Infas. *Mobilität in Deutschland (MiD) 2008*. Bundesministeriums für Verkehr, Bau und Stadtentwicklung, Germany, 2008.
5. Bundesamt, S. Berufspendler: Infrastruktur wichtiger als Benzinpreis. In *STATmagazin: Arbeitsmarkt*, Wiesbaden, 2014.
6. Legrain, A., N. Eluru, and A. M. El-Geneidy. Am Stressed, Must Travel: The Relationship between Mode Choice and Commuting Stress. *Transportation Research Part F: Traffic Psychology and Behaviour*, Vol. 34, 2015, pp. 141–151. https://doi.org/10.1016/j.trf.2015.08.001.
7. Trommer, S., V. Kolarova, E. Fraedrich, L. Kröger, B. Kickhöfer, T. Kuhnimhof, B. Lenz, and P. Phleps. Autonomous Driving: The Impact of Vehicle Automation on Mobility Behaviour. The Institute for Mobility Research (ifmo), Germany, 2016.
8. Fraedrich, E., R. Cyganski, I. Wolf, and B. Lenz. User Perspective on Autonomous Driving. A Use-Case-Driven Study in Germany. *Arbeitsberichte. Gerographisches Institut, Humboldt-Universität zu Berlin*, Vol. 187, 2016.
9. SAE International. Taxonomy and Definitions for Terms Related to On-Road Motor Vehicle Automated Driving Systems, SAE J3016. SAE International, 2014.
10. Kröger, L., and B. Kickhöfer. Autonomous Car- and Ride-Sharing Systems: A Simulation-Based Evaluation of Various Supply Options for Different Regions. In *ITEA Annual Conference on Transportation Economics*, Barcelona, 2017.
11. Yap, M. D., G. Correia, and B. V. Arem. Valuation of Travel Attributes for Using Automated Vehicles as Egress Transport of Multimodal Train Trips. *Transportation Research Procedia*, Vol. 10, 2015, pp. 462–471.
12. Boston Consulting Group. *Self-Driving Vehicles, Robo-Taxis, and the Urban Mobility Revolution*, 2016. https://www.bcg.com/publications/2016/automotive-public-sector-self-driving-vehicles-robo-taxis-urban-mobility-revolution.aspx.
13. Ohnemus, M., and A. Perl. Shared Autonomous Vehicles: Catalyst of New Mobility for the Last Mile? *Built Environment* Vol. 42, No. 4, 2016, pp. 589–602. https://doi.org/10.2148/benv.42.4.589.
14Burns, L. D., W. C. Jordon, and B. A. Scarborough. *Transforming Personal Mobility*. Earth Institute, Columbia University, NY, 2013.
15. Organisation for Economic Co-operation and Development (OECD), International Transport Forum (ITF), and Corporate Partnership Board. *Urban Mobility System Upgrade. How Shared Self-Driving Cars Could Change City Traffic*. ITF, Paris, 2015.
16. Gruel, W., and J. M. Standford. Assessing the Long-Term Effects of Autonomous Vehicles: A Speculative Approach. *Transportation Research Procedia*, Vol. 13, 2016, pp. 18–29.
17. Bahamonde-Birke, F. J., B. Kickhöfer, D. Heinrichs, and T. Kuhnimhof. A Systemic View on Autonomous Vehicles: Policy Aspects for a Sustainable Transportation Planning. *Kuhmo-Nectar Conference of the International Transportation Economics Association*, Barcelona, Spain, 2017.
18. Hensher, D. A. Valuation of Travel Time Savings. In *A Handbook of Transport Economics*. (A. de Palma, R. Lindsey, E. Quinet, and R. Vickerman, eds.). Edward Elgar Publishing Limited, Cheltenham, 2011, pp. 135–159.
19. Jara-Diaz, S. R. Allocation and Valuation of Travel-Time Savings. In *Handbook of Transport Modelling* (D. A. Hensher , and K. J. Button, ed.). Elsevier Science Ltd, UK, 2000, pp. 363–379.
20. Shires, J. D., and G. C. D. Jong. An International Meta-Analysis of Values of Travel Time Savings. *Evaluation and Program Plannung* Vol. 32, No. 4, 2009, pp. 315–325.
21. Abrantes, P. A. L., and M. R. Wardman. Meta-Analysis of UK Values of Travel Time: An Update. *Transportation Research Part A: Policy and Practice*, Vol. 45, No. 1, 2011, pp. 1–17.
22. Mackie, P., M. Wardman, A. Fowkes, G. Whelan, J. Nellthorp, and J. Bates. Values of Travel Time Savings in the UK – Full Report. *ITS Working Papers 561-566*, 2003.
23. Rizzi, L. I., J. P. Limonado, and S. S. C. Steimetz. The Impact of Traffic Images on Travel Time Valuation in Stated-Preference Choice Experiments. *Transportmetrica*, Vol. 8, No. 6, 2012, pp. 427–442.
24. Gucwa, M. Mobility and Energy Impacts of Automated Cars. In 2014 *Automated Vehicle Symposium Proceedings-Final*. Volpe National Transportation System Centers, Cambridge, MA, USA, 2014, p. 19.
25. Childress, S., B. Nichols, B. Charlton, and S. Coe. Using an activity-based model to explore possible impacts of automated vehicles. *Proc., 94th Annual Meeting of the Transportation Research Board*, Washington, D.C., 2015.
26. Winter, K., O. Cats, K. Martens, and B. V. Arem. A Stated Choice Experiment on Mode Choice in an Era of Free-Floating Carsharing and Shared Autonomous Vehicles. *Proc., 96th Annual Meeting of the Transportation Research Board*, Washington, D.C., 2016.
27. Krueger, R., T. H. Rashidi, and J. M. Rose. Preferences for shared autonomous vehicles. *Transportation Research Part C*, Vol.69, 2016, pp.343–355.
28. Becker, F., and K. W. Axhausen. Predicting the Use of Automated Vehicles. [First results from the pilot survey]. In *17th Swiss Transport Research Conference (STRC)*, Institute for Transport Planning and Systems. Monte Verità , Ascona, 2017.

29. Haboucha, C. J., R. Ishaq, and Y. Shiftan. User Preferences Regarding Autonomous Vehicles. *Transportation Research Part C*, Vol. 78, 2017, pp. 37–49.
30. Rose, J., M. Bliemer, D. Hensher, and A. Collins. Designing Efficient Stated Choice Experiments Involving Respondent Based Reference Alternatives. Working paper. Institute of Transport and Logistics Studies, University of Sydney, Australia, 2005.
31. Axhausen, K., I. Ehreke, A. Glemser, S. Hess, C. Jödden, K. Nagel, A. Sauer, and C. Weis. Ermittlung von Bewertungsansätzen für Reisezeiten und Zuverlässigkeit auf der Basis eines Modells für modale Verlagerungen im nicht-gewerblichen und gewerblichen Personenverkehr für die Bundesverkehrswegeplanung, 2014.
32. Allgemeiner Deutscher Automobil-Club. ADAC Autokosten Frühjahr/Sommer 2017, 2017. https://www.motor-talk.de/forum/aktion/Attachment.html?attachmentId=755768.
33. ChoiceMetrics. Ngene 1.1.1 User Manual & Reference Guide. ChoiceMetrics, Australia, 2012.
34. Bliemer, M. C. J., and J. M. Rose. Designing Stated Choice Experiments: State-of-the-Art. In *The Expanding Sphere of Travel Behaviour Research: Selected Papers from the 11th International Conference on Travel Behaviour Research* (R. Kitamura, T. Yoshii, and T. Yamamoto, eds.), Emerald, Bingley, UK, 2009, pp. 439–537.
35. Orme, B. K. Sawtooth Software's Lighthouse Studio (Version 9.1) [Computer software], 2017.
36. Hensher, D. A., and W. H. Greene. *The Mixed Logit Model: The State of Practice*. Working Paper ITS-WP-02-01, 2002. http://ws.econ.usyd.edu.au/itls/wp-archive/ITLS-WP-02-01.pdf.
37. McFadden, D. L. Conditional Logit Analysis of Qualitative Choice Behavior. In *Frontiers in Econometrics (P. Zarmebka ,ed.)*. Academic Press, New York, 1974, pp. 105–142.
38. Algers, S., P. Bergström, M. Dahlberg, and J. L. Dillen. *Mixed Logit Estimation of the Value of Travel Time*, No. 98-08-26, 1998.
39. Train, K. E. *Discrete Choice Methods with Simulation*. Cambridge University Press, NY, 2002.
40. Revelt, D., and K. Train. Mixed Logit with Repeated Choices: Households' Choices of Appliance Efficiency Level. *Review of Economics and Statistics*, Vol. 80, No. 4, 1997, pp. 647–657.
41. Box, G. E. P., and D. R. Cox. An analysis of transformations. *Journal of the Royal Statistical Society Series B*, Vol. 26, No. 2, 1964, pp. 211–252.
42. Bielaire, M. A Free Package for the Estimation of Discrete Choice Models. In *3rd Swiss Transportation Research Conference* Ascona, Switzerland, 2003.
43. Hess, S., K. E. Train, and J. W. Polak. On the use of Modified Latin Hybercube Sampling (MLHS) Method in the Estimation of a Mixed Logit Model for Vehicle Choice. *Transportation Research Part B*, Vol. 40, 2006, pp. 147–163.

The Standing Committee on Transportation Planning Applications (ADB50) peer-reviewed this paper (18-02795).

Article

TRR
JOURNAL OF THE TRANSPORTATION RESEARCH BOARD

Transportation Research Record
2018, Vol. 2672(46) 21–30

DOI: 10.1177/0361198118758632
journals.sagepub.com/home/trr
SAGE

Evaluating the Ability of Transit Direct Ridership Models to Forecast Medium-Term Ridership Changes: Evidence from San Francisco

Richard A. Mucci[1] and Gregory D. Erhardt[1]

Abstract

Transit direct ridership models (DRMs) are commonly used both for descriptive analysis and for forecasting, but are rarely evaluated for their ability to predict beyond the estimation data set. This research does so, using two DRMs estimated for rail and bus ridership in San Francisco. The models are estimated from 2009 data, applied to predict 2016 conditions, and compared to actual 2016 ridership. Over this period in San Francisco, observed rail ridership increased by 9% whereas observed bus ridership decreased by 13%. The results show that the models predict 2016 ridership about as well as that for 2009. The rail model correctly predicts the direction of change, but underestimates the magnitude of change. The bus model predicts the direction of change incorrectly, with a predicted 2% increase. A series of sensitivity tests are conducted to better understand the factors driving the ridership changes. These tests produce reasonable rail sensitivities, but reveal that the bus model is too sensitive to frequency, potentially because of the difficulty of estimating the coefficient from cross-sectional data when high-frequency transit also occurs in high-density locations. As the travel forecasting community increases its focus on empirically evaluating forecasts beyond a base year, DRMs must be a part of that.

Transit direct ridership models (DRMs), sometimes referred to as direct demand models, are often used for transit planning applications (*1*). A DRM is a sketch planning tool that reflects the notion that ridership is the sum of riders boarding at each stop. Rather than estimating trips on all modes and then applying a mode choice, as is done in the four-step modeling paradigm, DRMs estimate transit ridership "directly" as a function of station or stop environments and transit services. Essentially each station becomes a node and a unit of analysis, which provides fine-grained spatial analysis of factors influencing ridership.

A hybrid between a DRM and a four-step model is the Federal Transit Authority's (FTA's) Simplified Trips on Stops (STOPS) model (*2*), which estimates rail and bus rapid transit (BRT) project ridership. It is similar to a DRM, in that it predicts only the ridership for the rail or BRT routes, but it considers more information than most DRMs. The additional information includes the Census journey-to-work flows, and highway level-of-service skims from a regional travel model.

DRMs have been developed for both rail and bus transit, and used for a variety of applications. The models are used both as descriptive tools to understand the factors that influence ridership, and as predictive tools to forecast the ridership at new stations or in response to changes in service or other attributes.

For example, Cervero et al. (*3*) estimate a DRM of BRT ridership in the Los Angeles area. They argue for the application of DRMs as a complement to four-step models, to generate first-cut ridership estimates, and to conduct sensitivity tests of key variables. Pulugurth and Agurla (*4*) estimate DRMs using different spatial modeling methods in Charlotte, North Carolina. They find that using a spatial weight method does not yield better results over a spatial proximity method, a buffer. Kepaptsoglou et al. (*5*) predict ridership for a new light rail transit system in Cyprus using a DRM. The transit system has not been built yet, so there was not any comparison to observed ridership.

Whereas the goal of the previous papers was to forecast transit ridership, the following papers explore the effects of various factors on transit ridership. Lee et al. (*6*) study the relationships between heavy and light rail ridership and socioeconomic characteristics in South Korea. They find that heavy rail ridership is correlated with the density of businesses and households, whereas light rail ridership is correlated with the density of economically active population and businesses. Liu

[1]Department of Civil Engineering, University of Kentucky Lexington, KY

Corresponding Author:
Address correspondence to Alex Mucci: Alex.Mucci@uky.edu

et al. (7) estimate a DRM of rail ridership in Maryland to suggest improvements that will boost ridership. Holmgren (8) determines what factors are affecting transit ridership in Sweden and does this by estimating an econometric first-difference model. Estupian and Rodriguez (9) estimate a DRM of Bogota's BRT ridership to quantify the effect that the built environment has, in particular environmental supports for walking. Dill et al. (10) describe the effects that transit service characteristics and urban form have on local bus ridership. They find that sociodemographic characteristics seem to have a larger effect on ridership in large urban areas rather than small urban areas.

In nearly all of the above cases, the DRM is estimated from cross-sectional ridership data using ordinary least squares (OLS) regression. An exception is from Kerkman et al. (11) who use data from both 2012 and 2013, during which significant service changes occurred. They estimate cross-sectional models and compare those models to a fixed-effects panel data regression model in which the dependent variable is the change in ridership from 2012 to 2013. They find notably different sensitivities between the two models, with the panel data models showing the elasticity of ridership with respect to changes in stop frequency as about half of what the cross-sectional models show. They suggest that the panel data models do a better job of mitigating endogeneity when transit agencies tend to run higher frequency service in locations with higher potential demand.

Whereas DRMs are routinely promoted as tools for forecasting (2, 4), they are rarely evaluated for their ability to forecast. In fact, many of the published DRMs evaluate the models exclusively on the goodness of fit against the estimation data set, without any validation against independent data in either location or time (2–9). An exception to this is Upchurch and Kuby (12) in which a DRM is estimated using data in nine (other) cities, applied to predict the ridership on a new light rail line in Phoenix, and compared to actual ridership. The comparison allowed the authors to understand how well the model could be applied to a new data set, and to understand what caused over- and underpredictions.

Beyond the efforts of Upchurch and Kuby, there remains a sparsity of evidence on the ability of DRMs to forecast changes in transit ridership. This research contributes to filling that gap by estimating both a rail and bus DRM using 2009 data, applying that model to forecast 2016 ridership, and comparing the forecasts to actual 2016 conditions. It does this using the example of the San Francisco Municipal (MUNI) light rail and bus system in San Francisco, California. The analysis goes on to examine the magnitude of factors affecting the change in modeled ridership between these two years. Together, this allows us to evaluate how well the models perform in a predictive way, what they are able to capture in drivers of transit ridership trends, and what they may be missing.

The remainder of this paper is structured as follows. The next section describes the context of the San Francisco transit system. The third section explains the data sources and data processing. The fourth section presents the model estimation results for both the rail and bus models. The fifth section shows the application of the models to predict 2016 conditions. The sixth section examines the factors driving the change in modeled ridership. Finally, conclusions and next steps are considered.

Context

The nine-county San Francisco Bay area covers approximately 7,000 square miles with a population of 7.5 million residents. San Francisco is one of the core counties and has a population of 850,000 residents. The major transit systems operating in San Francisco are MUNI, Bay Area Rapid Transit (BART), and Caltrain. The system of interest for this paper is MUNI, and in particular the bus and light rail systems.

The MUNI system experienced significant service cuts in 2010 because of budget constraints. Since then, there have been incremental service improvements, which have accelerated from about 2014 as the agency implements its MUNI Forward program (13), focused on deploying Rapid (a skip-stop service with associated operational improvements) bus routes in high-use corridors.

During the past decade transit ridership in the area has experienced diverging growth, with bus ridership declining and rail ridership growing significantly (14). Our data show that, between 2009 and 2016, Muni rail ridership increased from 147,500 to 161,400, whereas MUNI bus ridership decreased from 515,000 to 450,000. The employment and population in San Francisco has increased significantly, so it would be natural to expect a corresponding growth in transit ridership. By quantifying the factors that contribute to ridership changes, and comparing the total observed and modeled change, this research provides insight into both the accuracy of the model and explanations for these divergent trends.

Data

DRMs estimate boardings or exits or both at a stop for defined periods of time, daily for this paper, as a function of variables related to the stop. This paper uses the standards previous research has set for DRMs with slight modifications to improve the models performance in San Francisco. The first modification is that the data are aggregated to the (directional) route-stop level rather than the stop level. That means that if two different transit routes stop at the same location, they are treated as separate observations. When the locational attributes of each record are the same, this allows the service attributes to be tracked separately. Second, bus stop-area data are tabulated for tenth-mile buffers instead of the more common quarter-mile buffers. This reflects the high density of stops in the network and avoids excessive overlap

Table 1. Significant Variables in Previous Rail and Bus DRMs

Service attributes	Location of stop attributes	Bus stop or site attributes
Presence of dedicated-lane services (0 or 1)	Population density	Presence of bus benches (0 or 1)
Run speed (actual)	Employment density	Presence of a passenger information system (0 or 1)
Reliability	Total urban density	Presence of a bus-stop shelter or canopy (0 or 1)
Bus-stop ridership (log) (dependent variable)	Distance to nearest rapid transit stop	Distance to urban center
Fare	Income	Bus terminus (0 or 1) (Start and end of a line)
Service miles (scheduled)	Competitive bus-stops	
Dwell time		
Crowded hours		
Secondary attributes		
Number of rear door boardings	Percentage of population that are elderly	Transfer stop (0 or 1) (2 or more routes can be accessed)
Number of daily connecting rail-transit lines	Land use (residential/agriculture/etc.)	Presence of a park-and-ride lot (0 or 1)
Supply demand match index (0–1)	Street connectivity (# of nodes divided by # of links)	Number of park-and-ride spaces
Stop frequency (log)		Presence of BRT branding or logo at the stop (0 or 1)

between adjacent buffers. Rail stations use the larger 0.25 mile buffer. Both were tested with a range of buffer sizes.

Previous research was used to determine what other data should be collected and included in the model (*5–9*). Table 1 shows an overview of significant influencers of transit ridership with previous models. These variables were used as a starting point to guide the data collection process.

All variables with log+1 in parenthesis are the log-transformed version of the data. The transformation includes taking the natural log of the data plus one. The plus one is to avoid errors when the data are 0. A sample calculation can be found below:

$$\text{Data}_{\text{log-transformed}} = \ln\left(1 + \text{Data}_{\text{original}}\right).$$

Table 2 shows a list of the variables considered in this research, along with their calculation method. The remainder of this section describes the data and how it is processed in detail.

Bus Ridership (Dependent Variable)

A portion of the San Francisco MUNI bus fleet is equipped with automated vehicle location/automated passenger counter (AVL/APC) technology. The AVL equipment records the latitude and longitude along with a timestamp every time a vehicle arrives or departs a stop. Buses are randomly assigned the AVL/APC equipment at the depot each day, but over several days all routes have been observed. The data are scaled up to cover the entire system by applying the ratio of total trips over observed trips to the observed data using a previously developed process (*14*). In addition to ridership, the data provides other performance metrics, including runtime, speed, on-time performance, and crowding measures. This analysis considers average weekday conditions (excluding holidays) for the fourth quarter of 2009 and 2016.

Rail Ridership (Dependent Variable)

Light rail cars are not equipped with APCs. Instead the San Francisco Municipal Transportation Authority (SFMTA), relies on occasional manual counts. They provided boarding and alighting counts at the station level for average weekday conditions in both 2009 and 2016. Because of the limited ridership data in 2009, some routes (the J, KT, and L) were supplemented with 2008 counts.

General Transit Feed Specification

GTFS (*15*) allows transit operators to publish their schedules in a standard format. Only the current schedule is published, but when a new version is published, the old is archived (*16*), so the differences can be used to systematically identify transit service changes. The GTFS provides service attributes for

Table 2. Variables and Data Sources

Variable	Variable name	Description	Data source	Calculation method
Ridership (dependent variable)				
Bus/rail route-stop ridership (Log+1)*	LOG_RIDERS	Daily average number of bus/rail passengers boarding and alighting a bus at each specific route-stop	MUNI (bus operator)	Automated passenger counter and General Transit Feed Specification (GTFS) data fusion
Potential demand (independent variables)				
Employment Development Department (EDD) employment (Log+1)	EDD_EMP LOG	Total employment within walking distance of a stop	EDD (San Francisco County Transportation Authority, SFCTA)	Buffer aggregation
Longitudinal Employer–Household Dynamics employment density	EMP WAC DEN	Employment per acre within walking distance of a route-stop	Longitudinal employer-household dynamics	Buffer aggregation
Income (2009 dollars)	SHR INCOME 100P [a]	Distribution of income for a census tract that a route-stops falls within. Incomes are in 2009 dollars for both 2009 and 2016	American Community Survey (ACS)	Spatial join
Housing density (Log+1)	HOUSING DEN LOG	Housing units per acre within walking distance of a route-stop	Census and Housing Inventory (SFCTA)	Buffer aggregation
Population density	POP DEN	Population per acre within walking distance of a route-stop	Census	Buffer aggregation
Population	TOTAL POP	Total population for a census tract that a route-stop falls within	ACS	Buffer aggregation
Share of households with zero vehicles	SHR HH 0VEH	Share of households with zero vehicles for a census tract that a route-stop falls within	ACS	Spatial join
On-street parking cost (Log +1) (2014 dollars)	PARK HOURLY AVG ON LOG	Hourly cost of on-street parking in 2014 dollars	SFCTA	Buffer aggregation
Transit supply (independent variables)				
Frequency (Log+1)	FREQ S LOG	Average number of buses per hour scheduled to serve each route-stop on a typical weekday	GTFS	Freq = 1/ (Headway*60)
Competitive stops	COMP STOPS	The number of stops within walking distance of a route-stop	GTFS	Buffer aggregation
Bus terminus	EOL SOL	Whether or not a route-stop is the start or end of a route	GTFS	na
Transbay Terminal (bus station)	TRANSBAY	Whether or not a route-stop is within walking distance of the Transbay Terminal	Google Earth	Buffer aggregation
Reliability	ONTIME5	The on-time share of a route-stop, specified as either 1 min before or 5 min after the scheduled time	GTFS	N/A
BART ridership	AVG_BART_LOG	The average number of BART passengers boarding and alighting within walking distance of a route-stop	BART Origin-Destination (OD) matrices	Buffer aggregation
Route configuration	LIMITED/EXPRESS	A variable to separate out the limited/rapid and express routes from the rest of the bus fleet	GTFS	na
Bus/rail ridership[b]	MUNI BUS AVG/ MUNI_RAIL_AVG	The average number of bus/rail passengers boarding and alighting within walking distance of a route-stop.	MUNI (bus operator)	Buffer aggregation

[a]This is one category of income and includes the share of households with an income of over $100,000.
[b]For the bus model the bus ridership is the dependent variable with the rail ridership as an independent variable, and vice versa for the rail model.

both the rail and bus models, such as scheduled headways, and is used as trip totals when the APC data are scaled up.

Employment Development Department

EDD data for the fourth quarter of 2009 and 2016 were used as the measure of employment. The data are segmented by industry, and are spatially detailed. Longitudinal employer–household dynamics (*17*) was considered, but not used because it is currently only available through 2014.

Housing Inventory and Decennial Census

The San Francisco Planning Department provided housing inventory data, which tracks the completion of housing developments in the city with the address, completion date, and net change in housing units. The housing inventory was used to pivot from the 2010 Census blocks to obtain an estimate of the housing stock in each year.

American Community Survey

To test the possible effect of demographics and socio-economic characteristics, we used ACS data (*18*). For the 2009 data set, we used the 5-year estimates, from 2005 to 2009. For 2016, we used the 5-year estimates from 2011 to 2015, the most recently available. The 5-year estimates are more suitable for our purpose because they provide more spatial detail than the 1-year estimates. In tracking demographic changes, we did not use a stop-buffering approach, but instead tagged each stop with the attributes of the census tract in which it lies. The income variable for 2015 had to be adjusted to 2009 dollars, to account for inflation; this means that the share of households above \$100,000 in 2009 dollars is the share above \$115,000 in 2015 dollars. The 2009 income variable is kept at \$100,000, as it is already in 2009 dollars. The \$115,000 was found by multiplying \$100,000, the 2009 original amount, by the ratio of the 2016 consumer price index (CPI) over the 2009 CPI.

BART Monthly Entry and Exit Matrices

The BART system is a heavy rail system serving four counties in the San Francisco Bay area. Commuters are a large share of the users because the system provides an alternative to the heavily congested Bay Bridge. BART serves as a potential competitor to MUNI for certain trip interchanges within San Francisco, but also serves as a complement because many regional BART trips transfer to MUNI for the "last-mile".

BART uses distance-based fares, so passengers have their tickets read on both entering and exiting the system. Knowing the number of entries and exits for each station allows BART to publish monthly matrices showing the number of trips to and from each station (*19*). There are nine BART stations in San Francisco that are of interest: Embarcadero, Montgomery, Powell, Civic Center, 16th, 24th, Glen Park, Balboa Park, and the Daly City station, the latter being included because of a few of the buffers reach across the county line.

Hourly On-Street Parking Cost

Parking data was provided by SFCTA and came in the form of hourly on-street parking prices throughout the day. The prices were averaged to obtain an average hourly price for a meter on a typical weekday. The data was then attached to micro-analysis zones to give the data a spatial component and then assigned to a stop by using a buffer around the stop. These data are only available for 2014, and are not collected annually. The hourly prices are kept in 2014 dollars because that is the year the data are observed. As the data was only available for 1 year the same prices are used for 2009 and 2016. Thus the parking prices can only be used to look at the relationship between them and transit ridership, and not how they influence changes in transit ridership.

Model Estimation

This section presents the model estimation results for the bus and rail models.

Bus Model

A 2009 base year model was estimated for each of the modes. The Python Statsmodels package (*20*) was used to run OLS regression on the data. The dependent variable was the log-transformed average of the number of passengers boarding and alighting at each route-stop. The log transformation was made because of the ridership data being skewed toward 0. Table 3 shows the results of the 2009 bus model.

Each of the terms included in the model are significant at the 95% level or better. For those variables that are log transformed, the coefficients can be interpreted directly as an elasticity, as the left-hand side is also log transformed. Attributes that are measured based on the area (total employment, housing density, etc.) are calculated within a tenth-mile buffer of the stop. Attributes associated with the service itself (frequency, on-time performance, etc.) are measured at the route-stop itself. Demographic measures from the ACS (income shares) are measured based on the census tract that contains the stop.

The intercept can be interpreted as scaling the line of fit, by adjusting the slope. The scaling value is found by taking the exponent of the intercept, which is found to be 0.44.

The positive employment coefficient is expected, indicating that ridership grows with increasing employment. We also tested employment segmented by different industry categories, including retail, hotel and restaurant, and education and health, but found that these specifications did not result in logical coefficients.

Table 3. 2009 MUNI Bus Direct Ridership Model Results

Variable	Variable name	Coefficients	*T*-statistics	Observations
Intercept		−0.8151	−6.25	6261
Potential demand (independent variables)				R^2
EDD employment (Log+1)	EDD_EMP_LOG	0.1337	15	0.515
Housing density (Log+1)	HOUSING_DEN_LOG	0.1056	8.45	
High income households (2009 dollars)	SHR_INCOME_100P	−1.2371	−14.71	
On-street parking cost (2014 dollars)	PARK_HOURLY_AVG_ON_LOG	0.0231	3.44	
BART ridership (Log+1)	AVG_BART_LOG	0.059	6.87	
Rail ridership	MUNI_RAIL_AVG	7.40E−05	3.04	
Transbay Terminal (bus station)	TRANSBAY	0.8029	3.72	
Transit supply (independent variables)				
Frequency (Log+1)	FREQ_S_LOG	2.8359	57.78	
Reliability	ONTIME5	0.5476	5.96	
Close route-stops	CLOSE_STOP	−1.2928	−25.80	
Bus terminus	EOL_SOL	0.7281	12.75	
Limited route configuration	LIMITED	−1.2006	−18.35	
Express Route Configuration	EXPRESS	−1.911	−38.63	

Higher frequency service (measured in average vehicles per hour) is associated with higher ridership. This relationship is logical, but the model appears to be quite sensitive to frequency, with a 1% increase in frequency associated with a 2.8% increase in ridership. For comparison, other studies have found this elasticity to be within the range of 0.1 to 1.04 (*21*). This high sensitivity may occur because of endogeneity with the dependent variable, or because of collinearity with other descriptive variables. Either way, the highest frequency service, the highest employment densities, and the highest ridership all occur in downtown San Francisco, making it difficult for any cross-sectional model to parse out the differences. Headway, a linear treatment of frequency, and several piece-wise linear specifications were also tested, but with no better results.

Ridership is higher at the start and end of lines, in areas with denser housing, and in areas with higher parking costs, all of which we would expect.

Stops in areas with a higher share of households earning $100,000 or more per year tend to have fewer riders per transit stop. This may be because wealthier households have more options to take other modes. It is also noteworthy because virtually all of the net growth in households over the 2009 through 2016 period has been through an increase in households earning $100,000 or more per year (*14*).

We measure reliability as the share of buses on that route arriving at that stop on-time, where on-time is defined as no more than 1 min early and no more than 5 min late (*22*). This is of particular interest, as reliability has been an important focus of MUNIs operational efforts in recent years.

We measure BART ridership as the log-transformed Bart ridership calculated as the average of BART boardings and alightings within the buffer area. The positive coefficient indicates net complementarity between BART and MUNI, with transfers between the two. Boardings and alightings on MUNI light rail stops within the buffer area also have a positive and significant coefficient. Stops near the Transbay Bus Terminal, which serves AC Transit commuter buses from the East Bay, also have higher ridership, likely because of a similar transfer effect.

CLOSE_STOP is a dummy variable representing whether or not the previous stop on the same route is within 0.2 miles. It is included to account for competition with other stops on the same route, and the negative coefficient indicates that closely spaced stops have lower ridership per stop, as we would expect.

Limited and express routes tend to have fewer riders per stop, which may be associated with relatively worse off-peak service.

The overall R^2 of the model is 0.515 relative to a constants-only model.

Rail Model

The rail model used the same unit of observation, route-stop. There are fewer observations because there are significantly less rail stops than bus stops within San Francisco. Table 4 shows the estimation results for this model.

The rail model specification broadly follows that of the bus model, although it is simplified both as some insignificant variables drop out of the model, and because we also lack some of the operational measures, such as on time performance, that are available through the APC data on the buses. Station-area attributes for the rail model are tabulated using a 0.25 mile buffer, reflecting a higher willingness to walk to rail and longer distances between the stations. Tenth

Table 4. 2009 MUNI Rail Direct Ridership Model Results

Variable	Variable name	Coefficients	*T*-statistics	Observations
Intercept		2.1822	−6.245	315
Potential demand (independent variables)				R^2
EDD employment (Log)	EDD_EMP_LOG	0.1607	4.29	0.482
Housing density (Log)	HOUSING_DEN_LOG	0.2755	4.349	
BART ridership (Log)	AVG_BART_LOG	−0.0652	−2.935	
Bus ridership	MUNI_BUS_AVG	6.15×10^{-5}	4.452	
Transit supply (independent variables)				
Frequency (Log)	FREQ_S_LOG	0.3042	1.686	
Close stop	CLOSE_STOP	−0.3706	−3.18	
Bus terminus	EOL_SOL	0.3129	1.89	
Route J route-stops	J	−0.6166	−5.158	

and third mile buffers were also tested and rejected. All variables included in the model are significant at the 90% level.

The intercept can be interpreted as scaling the line of fit, by adjusting the slope. The scaling value is found by taking the exponent of the intercept, which is found to be 8.87.

The coefficient on the log-transformed employment is 0.16, which is slightly higher than what is found in the bus model. Specifications that segmented the employment by industry were unsatisfactory. Housing density is also positive and significant.

The log of transit frequency in the rail model is 0.3, and significant at the 90% level. This is much lower than what is found in the bus model, and may reflect that there is less variation in frequency between the more sparse stops.

Unlike with the bus ridership, BART boardings in the buffer area are negatively correlated with MUNI rail ridership. This may reflect stronger competition and less complementarity between the two rail systems. The model does still show complementarity with MUNI bus boardings and alightings within the buffer area. That symmetry is encouraging.

As with the bus model, a nearby rail stop on the same route draws down the ridership at the stop in question. Both end of line and start of line stations tend to have higher ridership as expected.

A constant on the J line is negative and significant. Constants tested on each of the other lines were insignificant and left out of the model. The negative coefficient could be associated to the J line competing with BART more directly. Many of the J line stops are close to BART stations.

The overall R^2 of the model is 0.482 relative to a constants-only model.

Other Variables Tested

There were numerous variables included in the estimation file, but were not included in the final model. A few of importance are the share of households with zero vehicles and fares. The share of households with zero vehicles is potentially explained with other variables, such as housing density. However, fares only change 25 cents between 2009 and 2016. After accounting for inflation the change is only 6 cents, which was considered to be negligible.

Model Application

To evaluate the forecasting ability of the models, the 2009 bus and rail models were run using an independent data set, the 2016 data that was processed in the same manner as the 2009 data. In contrast to the model estimation, which limited the data set to route-stops with observed boardings, the models were applied to the full set of route-stops in San Francisco (a few outside the county were excluded). The focus of the examination focuses on how well the models predict change, more so than their cross-sectional fit.

Table 5 shows the system-level results. The 2009 bus model under-predicts total ridership by 11%, and the rail model under-predicts total ridership by 13%. Between 2009 and 2016, the data show that bus ridership decreases by 13%, whereas the model predicts a 2% increase. The data show rail ridership increasing by 9% over this period, whereas the model only predicts a 3% increase. For the rail model, the percent root mean square error (RMSE) is slightly better for the 2016 application than for the 2009 application. The opposite result is found for the bus model. The percent RMSEs are fairly large because of the disaggregate nature of a route-stop. The percent RMSE is given at two aggregated levels, stop and route. The general trend is that a better percent RMSE is found when more aggregated data are used. The average observed ridership for a bus route-stop is 70, a stop is 130, and a route is 5,500. For rail, the average observed ridership for a route-stop is 410, a stop is 530, and a route is 25,600.

Factors Affecting Change

To better understand the change in ridership between the two years, a series of sensitivity tests were conducted with the models. The tests focused on the subset of route-stops

Table 5. System Level Model Application Results

	Observed ridership	Modeled ridership	Difference	Difference	Route-stop RMSE	Stop RMSE	Route RMSE
Bus							
2009	515,059	459,602	−55,456	−11%	409%	369%	150%
2016	449,819	467,522	17,704	4%	424%	369%	72%
Change	−65,240	7,920					
% Change	−13%	2%					
Rail							
2009	147,470	128,042	−19,428	−13%	67%	82%	20%
2016	161,387	131,721	−29,666	−18%	60%	79%	21%
Change	13,917	3,679					
% Change	9%	3%					

Table 6. Factors Affecting the Change in Modeled Bus Ridership from 2009 to 2016

		Average values (without log)			Associated ridership change		
Variable	Coefficient	2009	2016	% Diff.	Absolute	%	Elasticity
Intercept	−0.8151						
Potential demand							
EDD employment (Log)	0.1337	1197	1434	20%	10,403	4%	0.19
Housing density (Log)	0.1056	384	396	3%	2,236	1%	0.26
On-street parking cost (Log) (2014 dollars)	0.0231	94	93	−1%	(173)	0%	0.10
High income households (2009 dollars)	−1.2371	0.36	0.38	6%	(11,245)	−4%	−0.75
BART ridership (Log)	0.059	282	294	4%	601	0%	0.05
MUNI rail ridership	7.40E−05	162	480	197%	11,959	5%	0.04
Transbay Terminal (bus station)	0.8029	0.002	0.008	278%	5,076	2%	0.01
Transit supply							
Frequency (Log)	2.8359	3.93	4.05	3%	37,224	15%	6.89
Reliability	0.5476	0.63	0.58	−8%	(11,091)	−4%	0.49
Close route-stops	−1.2928	0.90	0.90	0%	18,458	7%	45.84
Bus terminus	0.7281	0.05	0.05	1%	573	0%	0.16
Limited route configuration	−1.2006	0.02	0.03	18%	(589)	0%	−0.01
Express route configuration	−1.911	0.10	0.10	−2%	1,120	0%	−0.18
Total for route-stops present in both years—each term applied separately (Percentage change uses only ridership for 2009 stops present in 2016 too)					64,553	26%	
Total for route-stops present in both years—all terms applied together (Percentage change uses 2009 total ridership)					64,081	12%	
Total for route-stops dropped					(187,043)	−36%	
Total for route-stops added					130,882	26%	
System total					7,920	2%	

present in both 2009 and 2016, and each used 2009 data for all variables except for one isolation variable, for which 2016 data were substituted. This follows past work to understand the factors that drive changes in transit ridership (*23*), and provides a means for understanding the magnitude of change that can be attributed to that variable.

Table 6 shows the results of this exercise using the bus model. The coefficient for frequency was mentioned before

Table 7. Factors Affecting the Change in Modeled Rail Ridership from 2009 to 2016

Variable	Coefficient	Average values (without log)			Associated ridership change		Elasticity
		2009	2016	% Diff.	Absolute	%	
Intercept	2.1822						
Potential demand							
EDD employment (Log)	0.1607	11926	14933	25%	4,119	3%	0.13
Housing density (Log)	0.2755	1789	1985	11%	5,116	4%	0.37
BART ridership (Log)	−0.0652	1473	1348	−8%	−21	0%	0
Bus ridership	6.15E−05	4297	3977	−7%	−3,959	−3%	0.42
Transit supply							
Frequency (Log)	0.3042	4.5	4.83	7%	3,181	3%	0.34
Close route-stop	−0.3706	0.83	0.86	3%	−3,290	−3%	−0.83
Rail terminus	0.3129	0.08	0.08	−6%	−415	0%	0.05
Route J route-stops	−0.6166	0.16	0.16	0%	–	0%	0
Totals							
Total for route-stops present in both years—each term applied separately					4,733	4%	
Total for route-stops present in both years—all terms applied together					4,115	3%	
Total for route-stops dropped					−2,067	−1%	
Total for route-stops added					1,631	1%	
System total					3,679	3%	

to be high, and it shows up here again resulting in ridership increasing by 15%. San Francisco has experienced favorable economic growth in recent years and that is picked up by the high income variable increasing 6%, resulting in a 4% decrease of bus ridership. MUNI rail ridership has grown significantly and is found to complement bus ridership leading to an increase of 5%. Close stops are picking up recent changes in route configurations in which the MUNI forward program is reducing the number of stops along a route. Employment has grown resulting in bus ridership increasing 4%. The bus system overall has become more unreliable resulting in a decrease of 4% in ridership. The other variables do not experience much change and thus have a minimal effect on bus ridership. There is a decline in ridership associated with route-stops in 2009 but not 2016, and an increase associated with route-stops in 2016 but not 2009. Parking data was only available for 2014, so there was not any observed change in the data. This resulted in the on-street parking cost variable contributing no change in ridership.

Table 7 shows the sensitivity analysis for the rail model. The changes appear reasonable given the trends in the data, and are more modest than in the bus model. For the rail model, the bus ridership variable leads to a decrease in rail ridership, whereas the opposite effect is found for rail ridership in the bus model. This is because the rail system complements the bus system. Rail ridership has increased leading to an increase in bus ridership, whereas bus ridership has experienced a decline resulting in a decrease in bus ridership.

Conclusions

This research demonstrated the application of a rail and bus DRM estimated from 2009 data to predict 2016 conditions. This was for the MUNI transit system in San Francisco. The results show that the models predict 2016 ridership about as well as 2009. Even though this is encouraging (but not surprising for a mature system), we are also interested in how well the models predict the change in ridership over this period. The rail model correctly predicted the direction of change, but the modeled magnitude of change is about a third of the actual. The bus model predicted the wrong direction of change and is about 15% off from the true percentage change.

To better understand what factors drive the changes in transit ridership, a series of sensitivity tests were conducted to isolate the effects of each variable in the model. This analysis showed modest changes across several variables in the rail model. Even though these sensitivities appear reasonable, they do not capture all of the observed increase in rail ridership. The bus sensitivity tests show two variables that the model is very sensitive to: frequency and the presence of nearby stops on the same route. MUNI has increased service over this period, with a focus on implementing Rapid routes. The frequency increases and associated skip-stop service on high volume routes should result in a net increase in ridership, but the sensitivity to frequency in particular appears to be unreasonably high. We suspect that the cross-sectional estimation struggles to separate the effects of frequency versus employment density, parking cost, and other factors which are all highest in the downtown area. Future research should look into panel data regression or other methods that may be better able to separate these effects.

Beyond these specific terms, it appears that there are real-world trends over this period that the models are not able to capture. The divergent rail and bus ridership trajectories are noteworthy; it would be valuable to understand what is driving that divergence. One possibility is that travelers are making an increased use of transportation network companies (TNCs) as a substitute for bus trips. TNCs have experienced noteworthy growth with the inception of the mode in 2009, in San Francisco. The current research suggests that TNCs complement transit, but the research has not focused on San Francisco. Future work should explore this issue.

Finally, we have found the exercise of evaluating DRMs in forecasting mode to be insightful, and would encourage other DRM users to conduct this type of exercise. Increasingly, travel forecasters are asked to evaluate how well their models predict, and DRMs are not exempt. A natural extension of this research would be to evaluate the ridership predictions of similar models, such as the STOPS model mentioned in the introduction. The forecasted ridership of the aforementioned model can be compared to observed ridership similar to how this paper evaluates DRM predictions.

Acknowledgments

This work was funded, in part, by the San Francisco County Transportation Authority (SFCTA). Staff at SFCTA and the San Francisco Municipal Transportation Agency supplied the data and provided valuable input and advice.

References

1. Cervero, R. *Alternative Approaches to Modeling the Travel-Demand Impacts of Smart Growth*. University of California Transportation Center, 2008.
2. RSG. *User Guide: STOPS Simplified Trips-on-Project Software*. Federal Transit Administration, 2015.
3. Cervero, R., J. Murakami, and M. Miller. Direct Ridership Model of Bus Rapid Transit in Los Angeles County, California. *Transportation Research Record: Journal of the Transportation Research Board*, 2010. 2145: 1–7.
4. Pulugurtha, S. S., and M. Agurla. Assessment of Models to Estimate Bus-Stop Level Transit Ridership Using Spatial Modeling Methods. *Journal of Public Transportation*, Vol. 15, No. 1, 2012, p. 3.
5. Kepaptsoglou, K., A. Stathopoulos, and M. G. Karlaftis. Ridership Estimation of a New LRT System: Direct Demand Model Approach. *Journal of Transport Geography*, Vol. 58, 2017, pp. 146–156.
6. Lee, K. S., J. K. Eom, S. Y. You, J. H. Min, and K. Y. Yang. An Empirical Study on the Relationship Between Urban Railway Ridership and Socio-economic Characteristics. *Procedia Computer Science*, Vol. 52, 2015, pp. 106–112.
7. Liu, C., S. Erdogan, T. Ma, and F. W. Ducca. *How to Increase Rail Ridership in Maryland? Direct Ridership Models (DRM) for Policy Guidance*. Presented at 93rd Annual Meeting of the Transportation Research Board, Washington, D.C., 2014.
8. Holmgren, J. An Analysis of the Determinants of Local Public Transport Demand Focusing the Effects of Income Changes. *European Transport Research Review*, Vol. 5, No. 2, 2013, pp. 101–107.
9. Estupiñán, N., and D. A. Rodríguez. The Relationship Between Urban Form and Station Boardings for Bogotá's BRT. *Transportation Research Part A: Policy and Practice*, Vol. 42, No. 2, 2008, pp. 296–306.
10. Dill, J., M. Schlossberg, L. Ma, and C. Meyer. *Predicting Transit Ridership at the Stop Level: The Role of Service and Urban Form*. Presented at 92nd Annual Meeting of the Transportation Research Board, Washington, D.C., 2013.
11. Kerkman, Kasper, Karel Martens, and Henk Meurs. Factors Influencing Stop-Level Transit Ridership in Arnhem?Nijmegen City Region, Netherlands. *Transportation Research Record: Journal of the Transportation Research Board*, 2015. 2537: 23–32.
12. Upchurch, C., and M. Kuby. Evaluating Light Rail Sketch Planning: Actual Versus Predicted Station Boardings in Phoenix. *Transportation*, Vol. 41, No. 1, 2013, pp. 173–192.
13. San Francisco Municipal Transportation Agency. *MuniForward*. Muni Forward - Get On Board. http://muniforward.com. Accessed Jul. 15, 2017.
14. Erhardt, G. D., O. Lock, E. Arcaute, and M. Batty. A Big Data Mashing Tool for Measuring Transit System Performance. In *Seeing Cities Through Big Data* (P. (Vonu) Thakuriah, N. Tilahun, and M. Zellner, eds.), Springer International Publishing, Switzerland, 2017, pp. 257–278.
15. Google Transit APIs. *Overview|Static Transit*. Google Developers. https://developers.google.com/transit/gtfs/reference/. Accessed Jul. 22, 2017.
16. Mapzen. *Transitland*. Tranistland. https://transit.land/. Accessed Aug. 1, 2017.
17. US Census Bureau Center for Economic Studies. *US Census Bureau Center for Economic Studies Publications and Reports Page*. Longitudinal Employer-Household Dynamics. https://lehd.ces.census.gov/data/. Accessed Jun. 20, 2017.
18. U. S. C. Bureau. *American FactFinder: Download Center*. American Fact Finder. https://factfinder.census.gov/faces/nav/jsf/pages/download_center.xhtml. Accessed Jun. 10, 2017.
19. Bay Area Rapid Transit. *Ridership Reports|bart.gov*. Ridership Reports. https://www.bart.gov/about/reports/ridership. Accessed Aug. 1, 2017.
20. *StatsModels: Statistics in Python—statsmodels 0.8.0 documentation*. Statsmodels's Documentation. http://www.statsmodels.org/stable/index.html. Accessed Aug. 1, 2017.
21. Balcombe, R., R. Mackett, J. Preston, J. Shires, H. Titheridge, M. Wardman, and P. White. "The Demand for Public Transport: A Practical Guide," January 1, 2004. http://www.trpa.org/documents/rseis/New%20References%20for%20Final%20EIS/Balcombe%20et%20al%202004.pdf.
22. Kittelson & Associates, National Research Council (U.S.), Transit Cooperative Research Program, United States, and Transit Development Corporation. (eds.). *Transit Capacity and Quality of Service Manual*, 3rd ed. Transportation Research Board, Washington, D.C., 2013.
23. Erhardt, G. D. *Fusion of Large Continuously Collected Data Sources: Understanding Travel Demand Trends and Measuring Transport Project Impacts*. PhD thesis. University College London (UCL), 2016.

The Standing Committee on Transportation Planning Applications (ADB50) peer-reviewed this paper (18-06169).

Article

TRR
JOURNAL OF THE TRANSPORTATION RESEARCH BOARD

Transportation Research Record
2018, Vol. 2672(46) 31–40

DOI: 10.1177/0361198118756628
journals.sagepub.com/home/trr
SAGE

A Model of Ridesourcing Demand Generation and Distribution

Patrícia S. Lavieri[1], Felipe F. Dias[1], Natalia Ruiz Juri[2], James Kuhr[2], and Chandra R. Bhat[1]

Abstract

Ridesourcing has experienced exponential growth in recent years, yet its impact on individual travel are unclear and have not been adequately examined. Recently, an Austin-based ridesourcing company released a large dataset containing disaggregate trip-level information. In this research, we use this new dataset in tandem with several publicly available data sources to estimate two models: a spatially lagged multivariate count model, which is used to describe how many trips are generated in a specific zone on both weekdays and weekend days; and a fractional split model, which helps us identify the characteristics of zones that attract ridesourcing trips. Our results show spatial dependence in ridesourcing trips among proximally located zones, as well as correlation between weekday and weekend day trips originating in a zone. Another interesting finding is the identification of a possible substitution effect between ridesourcing and transit use for weekday trips. Moreover, our results suggest that different income segments in the population may use ridesourcing for different activity purposes. From a travel behavior researcher perspective, the results in this paper identify aggregate area-level variables impacting ridesourcing, which can guide future efforts to better understand the demand for ridesourcing as well as the demand for autonomous and connected vehicles.

The concept of ridesourcing, in which a loose collection of drivers offers transportation services to the wider public based on coordination through a centralized dispatch (generally a mobile application), has introduced a new form of modern mobility. This type of service is coordinated by transportation network companies (TNCs), such as Uber and Lyft, and has experienced exponential growth in recent years. For instance, it took Uber, the largest TNC, 6 years to reach the 1 billion trip milestone in 2015, but only an additional 6 months to reach the 2 billion trip milestone. One year later, the company exceeded 5 billion trips (*1*). Such a substantial growth reflects the convenience that ridesourcing offers to users by being a reliable, lower cost (compared with traditional taxi services), on-demand, and door-to-door transportation service that is requested and paid through smartphone apps. In addition, these services are also able to easily reach a large geographically dispersed market, because they are less regulated than taxi companies and rely on drivers' personal vehicles.

Even as ridesourcing has gained considerable traction and is widely prevalent today in most urban areas, its impact on individual travel are unclear and have not been adequately examined. This includes limited knowledge of which travel modes are being substituted by ridesourcing, what its potential impacts on vehicle ownership are, how ridesourcing may affect peak and off-peak travel patterns, and whether its convenience induces more or less travel. A main reason for the lack of studies on the impact of ridesourcing is the scarcity of publicly available data. However, to fill this void, some researchers have resorted to specialized user surveys or large-scale household travel surveys that collect limited information on ridesourcing preferences. For instance, Rayle et al. presented results on an exploratory survey conducted among ridesourcing users ($n = 321$) in San Francisco, and observed that, whereas ridesourcing is being used in lieu of taxis in many instances, at least half of the users replaced trips by modes other than taxi, including public transit and driving (*2*). Clewlow and Mishra studied the impact of ridesourcing on vehicle ownership, conducting a survey in seven major U.S. cities (*3*). In their study, 9% of the respondents indicated that they had made a decision to dispose of one or more household vehicles because of the availability of ridesourcing. The remaining 91% reported that they had not changed their vehicle ownership, with 16% indicating that they had no vehicles to begin with. Dias et al. studied determinants of the frequency of both ridesourcing and car-sharing trips, using a survey undertaken by the Puget Sound Regional Council

[1]Department of Civil, Architectural and Environmental Engineering, The University of Texas at Austin, Austin, TX
[2]Center for Transportation Research, Network Modeling Center, The University of Texas at Austin, Austin, TX

Corresponding Author:
Address correspondence to Chandra R. Bhat: bhat@mail.utexas.edu

(PSRC) and found that vehicle ownership affects these two modes differently (*4*). Households with vehicles are less likely to use car-sharing services; however, households with one or more vehicles and residing in a high-density location are more likely than their peers residing in low-density areas to use both ridesourcing and car-sharing services. Finally, Lavieri et al., using the same PSRC survey, analyzed the determinants of ridesourcing adoption (as one model component of a larger model system), and found that living in high residential density areas and having a tech-savvy lifestyle were important contributing factors that encouraged the adoption of ridesourcing (*5*). Overall, the transportation literature on ridesourcing is in its infancy, and studies based on actual trip records are virtually non-existent.

This paper contributes to the literature by modeling and analyzing the demand for ridesourcing based on an open-source database released by RideAustin, a non-profit TNC in Austin, Texas. Using 6 months of detailed trip data, including trip origin and destination location and corresponding time stamps, we develop a two-step modeling framework to investigate the generation and distribution of daily ridesourcing trips at the traffic analysis zone (TAZ) level (the RideAustin dataset does not provide user information and corresponding socio-demographic characteristics; therefore, our analysis is undertaken at the zonal level and relies on zonal demographics to infer ridesourcing demand characteristics). We use a spatial bivariate count model to analyze ridesourcing trip generation and inform our understanding of the characteristics of the demand for this service. The use of a spatial analysis technique is important because spatial dependencies in TAZ-level trip generation are likely to exist. Thus, for example, it is possible that individuals in close proximity (say in neighboring TAZs) will be influenced by each other's ridesourcing propensity through social interactions, leading to a lagged endogenous variable effect. In our analysis, we consider this type of spatial dependence through the use of a spatial lag model. Subsequently, we apply a fractional split distribution model to identify zonal characteristics that attract ridesourcing trips and to examine how far individuals are willing to travel by this mode. Examples of explanatory variables used in our analysis are zonal distributions of income, gender, race, age, population, and employment density. We also consider transit supply characteristics and land use information regarding presence of parks and universities.

The remainder of this paper is organized as follows. The next section provides a description of the multiple data sources used in our analysis and of characteristics of the area of our analysis. Then we describe the modeling frameworks, which is followed by the presentation of the results. The final section summarizes our findings.

Data

Several public datasets were compiled to undertake the analysis. The primary data source originated from RideAustin, a TNC operating in Austin, Texas. RideAustin entered the Austin ridesourcing market in 2016, shortly after Uber and Lyft shut down their operations in the city due to disputes over local regulations. The RideAustin data (*6*) provides trip-level information, including the location of trip origins and destinations, total trip length, and corresponding fare. To protect their clients' privacy, RideAustin added noise to the locations of the pickups and drop-offs. The dataset contains a total of 1,494,125 trips that occurred between June 4, 2016 and April 13, 2017. As ridership during the first few months was limited, our analysis only includes data from August 2016 through January 2017. Based on information provided by the Austin Department of Transportation, we estimate that, during that semester, RideAustin was responsible for one third of Austin's ridesourcing market share, suggesting high representativeness of the data. The trip information is supplemented using TAZ-level demographic data obtained from the Capital Area Metropolitan Planning Organization (CAMPO) website and planning toolkit, and the most recent Census estimates. General Transit Feed Specification (GTFS) data are used to estimate the characteristics of the transit system (*7*).

Data Preparation

Raw data, including trip origins and destinations, transit availability, land use, and demographics were mapped to the TAZs defined by CAMPO using GIS software. Given the sparseness of origins and destinations in the outskirts of the city, we chose to focus this study on trips that originate in Central Austin, in the region delimited by Highway 183 to the east and north, Highway 290 to the south, and Texas State Highway Loop 1 (MoPac) to the north. There are a total of 458 TAZs in the area of analysis. The Austin-Bergstrom International Airport is outside the area of interest, but it attracts a large number of trips, so it was modeled as a special external zone in the fractional split model; a second dummy TAZ was used as the destination of all the trips that end outside the study area.

The trip data processing involved calculating the average number of daily trips per origin, and the corresponding average daily split by destination. Separate values were computed for weekdays and weekends. Demographic variables by TAZ were computed using data from the most recent Census, while land-use variables were obtained from the CAMPO planning model. GTFS data were aggregated to generate metrics of transit accessibility, including transit stops per zone, and the average frequency of buses per stop for weekdays and weekends.

Data Description

This research models the average daily count of ridesourcing trips for weekdays and weekends. The spatial unit of analysis is a TAZ. Table 1 presents the descriptive statistics for all the variables used in the model, and the year when the corresponding data were collected.

The analysis of descriptive statistics shows a large dispersion in the number of trips generated per zone. Figure 1 illustrates the spatial distribution of trips on an average week and weekend day. There is a clear concentration of trips in central and denser areas on both types of days. During weekdays, trips are more concentrated in specific zones that contain universities, parks, or active night life. On average there are almost four times more trips generated on a weekend day than on a weekday. These observations are consistent with Rayle et al.'s results, which suggest, as in San Francisco, that ridesourcing in Austin too is used more for social and leisure activities than work-related activities (*2*). Indeed, Hampshire et al. recently conducted an online survey in Austin and identified the same pattern (*8*). On a related note, the average cost of a ridesourcing trip in our sample is US\$12.77.

The analysis of transit supply variables suggests that the distribution of transit in Austin is rather heterogeneous. The frequency of bus services is an average of 3.12 per hour during weekdays (an average headway of about 20 minutes) and an average of 1.64 per hour on weekend days (an average headway of about 37 minutes).

In terms of socio-demographics, there is again a large variation across zones in population density and employment density. The race/ethnicity and education variables indicate a predominantly white and highly educated population. There is a good distribution of individuals in the 18–60 age range. Households are small in size (average of less than two individuals), have a mean income of US\$48,000, and have high vehicle ownership rates (more than half of the sample has at least two vehicles per household). Finally, three variables considered in our model, but not presented in the table, are binary variables representing the presence of parks in a zone, the presence of the University of Texas (UT) territory in a zone, and an indicator of whether a zone is a central business district (CBD).

Analytic Framework

In this paper, we develop a two-step procedure to analyze ridesourcing trip generation and distribution between TAZs on weekdays and weekends. In the trip generation analysis, the average number of trips generated at each TAZ on an average weekday is modeled jointly with the average number of trips generated on an average weekend day. We utilize a spatial bivariate count model that takes into consideration the spatial dependence between TAZs as well as the correlation between the two types of days. Accounting for spatial dependency in trip generation models, as discussed earlier, is important because one can expect neighboring zones to present similar travel demand patterns, especially when considering that the delimitation of zone borders is usually made based on the transportation network and does not necessarily reflect differences in demand patterns. Of equal importance is to recognize that ridesourcing trip generation rates may vary between weekdays and weekends, but that common TAZ-level unobserved factors may influence the counts on both types of days. For instance, a zone with very limited parking (an unobserved variable in our analysis) is likely to be associated with high ridesourcing trip generation rates, both on weekdays and weekend days.

For the trip distribution analysis, we develop a fractional split distribution model that analyzes the fractions of trips originating from a zone that terminate in each destination zone. This analysis provides an understanding of factors that "pull" ridesourcing trips toward a zone. The next two sections provide an overview of both the generation and distribution models.

Spatial Multivariate Count Model

The spatial multivariate count model used in this paper is based on the model of Bhat et al. (*9*). There are two components to this model. The first part is the recasting of the basic count model and the second part is the spatial dependency formulation.

Count Model Recasting. The framework used here is based on a recasting of the basic count model as a special case of a generalized ordered-response (GOR) model, as proposed by Castro et al. (*10*). In this approach, the count is viewed as a result of a latent demand generation propensity that gets mapped into the observed trip counts through thresholds that are themselves functions of exogenous variables. This approach offers the advantage of accommodating over-dispersion and excess zeros, which is useful when modeling zones that do not produce any trips (for example, open areas) and zones that produce unusually high numbers of trips (for example, zones that have high density of bars and active night life).

Let q ($q = 1, 2, \ldots, Q$) be the index for the territorial unit of analysis (a "TAZ" in the current paper) and let s ($s = 1, 2, \ldots, S$) be the index for day type (weekday or weekend day in the current paper). Let y_{qs} be the index for the count of trips generated in a day type s in a TAZ q, and let m_{qs} be the actual observed count of trips in the day type s in the TAZ q. Next, consider that there is a TAZ-specific demand function that represents the propensity for trip generation on day type s. This propensity is not observed directly, and so may be represented by a latent (unobserved to the analyst) variable y^*_{qs}. Then, in the GOR notation, the latent propensity y^*_{qs} is written as a function of a $(K \times 1)$ vector of observed covariates $\boldsymbol{x}_q$ (excluding the constant) as

$$y^*_{qs} = \boldsymbol{\beta}'_s \boldsymbol{x}_q + \eta_{qs},\; y_{qs} = m_{qs} \Rightarrow \psi_{qs,m_{qs}-1} < y^*_{qs} < \psi_{qs,m_{qs}} \quad (1)$$

Table 1. Sample Descriptive Statistics (458 TAZs)

Variable	Min.	Max.	Mean	SD
Outcomes [2016]				
Number of trips in a weekday	0.00	125.00	8.62	11.18
Number of trips in a weekend day	1.00	420.00	31.45	19.89
Transit supply [2016]				
Number of bus stops	0.00	27.00	3.49	3.62
Frequency of buses in a weekday (bus per hour)	0.00	20.90	3.12	3.25
Frequency of buses in a weekend day (bus per hour)	0.00	13.70	1.64	1.66
Socio-demographic variables [2010]				
Population density (population per km^2)	0.00	59,257.67	4,603.24	6,167.92
Employment density (employment per km^2)	0.00	161,932.22	6,813.73	17,545.97
Employment density in retail sector (employment per km^2)	0.00	46,442.92	1,167.67	3,731.79
Race/ethnicity variables [2015]				
Proportion of White population	0.00	1.00	0.65	0.34
Proportion of Black and African American population	0.00	0.65	0.05	0.09
Proportion of Asian population	0.00	0.63	0.04	0.06
Proportion of other races	0.00	0.45	0.07	0.08
Educational attainment distribution [2015]				
Proportion of population 18 years and above with less than associate degree	0.00	1.00	0.31	0.26
Proportion of population 18 years and above with associate or bachelor's degree or higher	0.00	0.67	0.30	0.19
Proportion of population 18 years and above with graduate degree	0.00	0.64	0.20	0.15
Age distribution [2015]				
Proportion of population aged 17 years and below	0.00	0.48	0.13	0.11
Proportion of population aged 18–29 years	0.00	0.99	0.22	0.20
Proportion of population aged 30–39 years	0.00	0.44	0.15	0.10
Proportion of population aged 40–59 years	0.00	0.49	0.20	0.13
Proportion of population aged 60 years and above	0.00	0.44	0.11	0.10
Median household size [2010]	0.00	4.00	1.76	1.00
Median annual household income (US$) [2010]	0.00	248,200.00	48,812.00	48,049.00
Household vehicle Ownership [2010]				
Proportion of household with zero vehicles	0.00	0.08	0.02	0.01
Proportion of household with one vehicle	0.00	0.83	0.37	0.25
Proportion of household with two or more vehicles	0.00	0.95	0.61	0.28
Distance between centroids of census tracts (Km)	0.10	19.08	5.75	3.31

Note: SD = standard deviation.

In the above specification, β_s is a $(K \times 1)$ vector whose elements capture the effects of the x_q variable vector on latent demand propensity y^*_{qs}. Finally, η_{qs} captures TAZ-specific unobserved factors that increase or decrease the latent propensity for generating trips in a week or weekend day. The thresholds in Equation 1 take the following form:

$$\psi_{qs,m_{qs}} = \Phi^{-1}\left(e^{-\lambda_{qs}} \sum_{l=0}^{m_{qs}} \frac{\lambda_{qs}^{l}}{l!} \right) + \alpha_{s,m_{qs}}, \quad \lambda_{qs} = e^{\gamma_s' h_q}, \tag{2}$$

$$\alpha_{s,0} = 0, \ \alpha_{s m_{qs}} = \alpha_{s,L_s} \text{ if } m_{qs} > L_s,$$

where Φ^{-1} is the inverse function of the univariate cumulative standard normal, $\psi_{qs,-1} = -\infty \ \forall q, \ s, \ \alpha_{s,0} = 0 \ \forall s$ (this restriction is needed for identification, given the parameterization of the thresholds), h_q is a vector of exogenous variables (including a constant) associated with TAZ q (there can be common variables in h_q and x_q), γ_s is a corresponding coefficient vector to be estimated for day type s, and L_s is a pre-defined count level that is determined based on empirical testing and on the context under consideration. Note that thresholds are impacted by the TAZ characteristics so that the translation from the trip generation propensity into an actual number of trips generated may vary between two zones (based on the exogenous variables h_q) even if they have exactly the same x_q. As in the typical ordered-response framework, the values of $\alpha_{s,m_{qs}}$ should be such that the ordering condition on the thresholds $(-\infty < \psi_{qs,0} < \psi_{qs,1} < \psi_{qs,2} < ....)$ is satisfied. The presence of the $\alpha_{s,m_{qs}}$ terms provides flexibility to accommodate high- or low-probability masses for specific count outcomes without the need for cumbersome treatment using hurdle or zero-inflated mechanisms [see Castro et al. (*10*)]. If these terms are set to zero, and all elements of the vector β_s are also set to zero, the result is the traditional Poisson count model (*10*).

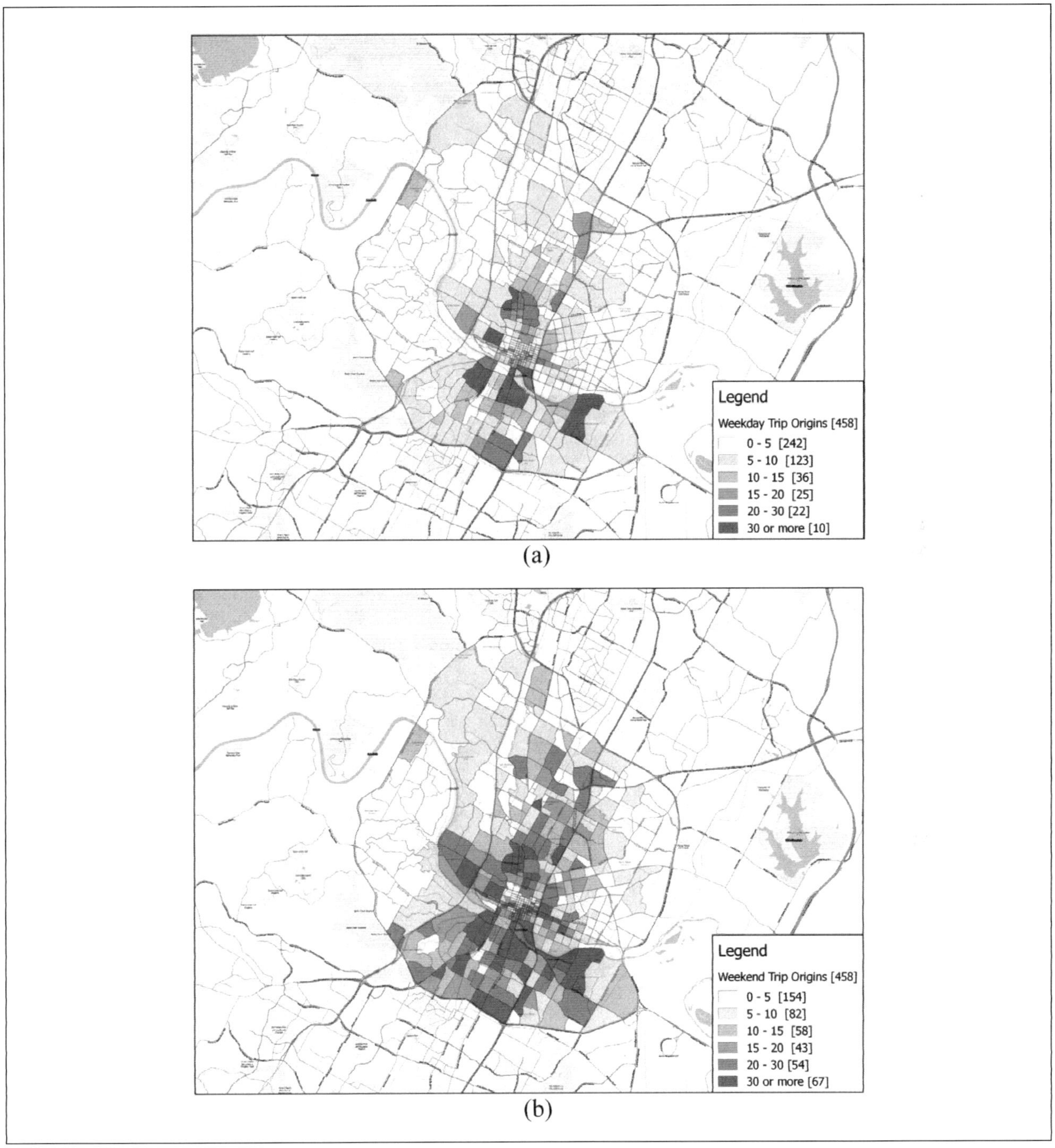

Figure 1. Spatial distribution of trips on (*a*) an average weekday and (*b*) on an average weekend day.

Spatial Component. The model adopted in this study considers spatial dependence across TAZs in observed covariates $\boldsymbol{x}_q$ vector as well as in the unmeasurable terms η_{qs}. To conserve on space, we refer the reader to Bhat et al. (*9*) for a complete explanation and formulation of the spatial structure of the model. In general terms, we have the usual distance-based spatial weight matrix (**W**), which indicates whether a pair of TAZ should be considered spatially dependent ($w_{qq'} > 0$). The trip generation propensity from a zone is influenced by exogenous variables specific to that zone, and the trip generation propensity from neighboring zones based on an autoregressive coefficient represented by δ_s. Here δ_s may be positive or negative $(-1 < \delta_s < 1 \; \forall \; s)$. In our model we adopt weight matrices based on functions of the distance between the center points of two TAZs. As we are analyzing the central area of Austin, we consider that spatial dependence may occur between zones that are up to 3 km apart. The final equation of the multivariate count model incorporating spatial dependence for a specific zone for a specific type of day is

$$\begin{aligned} y^*_{qs} &= \delta_s \sum_{q'=1}^{Q} w_{qq'} y^*_{q's} + \boldsymbol{b}'_s \boldsymbol{x}_q + \eta_{qs}, \\ y_{qs} &= m_{qs} \Rightarrow \psi_{qs,m_{qs}-1} < y^*_{qs} < \psi_{qs,m_{qs}} \end{aligned} \quad (3)$$

We consider the joint nature of the demand propensities across day types for each TAZ q by allowing the elements η_{qs} to be correlated across the two day types (s = 1, 2) for each TAZ q. A final important point to be noted here is that the spatial dependency in counts is generated through spatial "spillover" effects and spatial error correlation effects in the latent ridesourcing demand propensity, not through the localized TAZ-specific characteristics that impact the thresholds in the count model.

Fractional Split Distribution Model

To estimate the fractional split model, we use a quasi-likelihood estimation approach [see Sivakumar and Bhat (*11*) and Gourieroux et al. (*12*)]. Let f_{qj} be the fraction of the total trips that originate in zone q that terminate in zone j, such that $\sum_{j=1}^{J} f_{qj} = 1$. We can write the log-likelihood for the qj^{th} zone pairing as follows:

$$LL_{qj} = f_{qj} \log\left[G(\mu, z_{qj})\right] + (1 - f_{qj}) \log\left[1 - G(\mu, z_{qj})\right] \quad (4)$$

where $\boldsymbol{\mu}$ is a vector of coefficients to be estimated and z_{qj} is a vector of exogenous variables with characteristics of zones q and j. The function $G(.)$ takes the following logit functional form:

$$G(\mu, z_{qj}) = \frac{1}{1 + e^{-\mu' z_{qj}}} \quad (5)$$

Results

In this section we present and discuss the results for the trip generation and distribution models. We considered all the variables presented in Table 1 in our analysis. Several variables and functional forms (including logarithmic transformations) were tested to arrive at the final specification. The model estimation process was guided by prior research, intuitiveness, and parsimony considerations. A few variables that were only marginally statistically significant (i.e., not significant at the 0.05 level of significance) were retained in the final model specification because of their intuitive effects and potential to guide future research efforts.

Trip Generation by Day Type

As discussed previously, the count variable is viewed as a result of a latent demand generation propensity that gets mapped into the observed trip counts through thresholds that are themselves also functions of exogenous variables. The first half of Table 2 presents the results for the demand generation propensity. In the first row of results for the weekday trips model, we observe a positive effect of the variable representing the presence of UT in the zone. This positive effect indicates that UT is an area with a high intensity of ridesourcing trip origins during a typical weekday, presumably a combination of activity opportunities in the UT area and because students are a segment of the population more likely to use ridesourcing than other population segments. The effect of transit supply in Table 2 indicates the expected negative effect, suggesting that ridesourcing decreases as transit service improves. Another perspective is that ridesourcing tends to get used more in areas with relatively poor transit service. Areas with higher residential density and activity intensity, not surprisingly, have more originating ridesourcing trips, a finding supported by earlier studies [see (*3–5*)]. Interestingly, on weekdays, more ridesourcing trips are generated from high activity intensity zones than from high population density zones, while the reverse holds on weekend days. This suggests that ridesourcing is more used after an out-of-home activity on weekdays, and more used by individuals leaving home for an out-of-home activity on weekend days.

In terms of population characteristics, zones with higher proportions of white population present a lower propensity to generate ridesourcing trips, both during the weekday and the weekend day. The white population is historically associated with a higher use of the "drive alone" mode (*13, 14*) than other segments of the population, which may explain the negative signs on ridesourcing use. The proportion of young adults (18–29 years of age) in the zone contributes to an increase in the propensity of ridesourcing trips. This result corroborates findings from previous studies (*2, 3, 8*). The effect of the median household income of the zone is interesting. It shows that wealthier areas are associated with an increase in the weekday ridesourcing trips, but a decrease in weekend ridesourcing trips. The literature often suggests that ridesourcing users are in the high-income segments of the population (*3, 4*). Our results are not inconsistent with the previous literature, but suggest that there is heterogeneity in the income effect based on day of the week. Perhaps high-income individuals "buy" time on weekdays through ridesourcing (because they can work or relax rather than drive), whereas low-income individuals gravitate toward ridesourcing on weekend days because of relatively poor transit levels

Table 2. Estimation Results of the Trip Generation Count Model

Variables	Weekday	Weekend day
	Estimate (*t*-stat)	Estimate (*t*-stat)
Determinants of the latent demand generation propensity (y^*_{qs})		
Special land-use		
If all or part of the zone is occupied by The University of Texas	1.271 (2.34)	–
Transit supply		
Average frequency of buses in an average bus stop in the zone (bus per hour)	−0.023 (−1.75)	–
Residential density		
Population density (natural Logarithm of person per km^2)	0.436 (10.17)	0.753 (36.97)
Activity intensity		
Retail employment density (natural logarithm of retail jobs per km^2)	0.524 (11.74)	0.492 (14.58)
Population characteristics		
Proportion of White population	−3.644 (−10.18)	−0.488 (−11.13)
Proportion of population 18-29 years old	2.124 (5.64)	0.298 (1.94)
Proportion of population 30-49 years old	0.225 (1.95)	–
Median annual household income (divided by $10,000)	0.056 (3.77)	−0.073 (−3.58)
Proportion of households with 2 or more automobiles	–	−2.669 (−25.67)
Demand tipping points (threshold component)		
α_1	–	–
α_{10}	−0.240 (−2.96)	−0.350 (−2.66)
α_{20}	−0.474 (−2.55)	–
α_{25}	–	−0.661 (−5.48)
Determinants of the thresholds (γ vector elements)		
Constant	2.279 (50.92)	0.182 (6.39)
Population characteristics		
Proportion of male population	0.366 (14.08)	−0.345 (−6.75)
Households characteristics		
Proportion of households with two or more automobiles	−1.487 (−18.70)	–
Special land-use		
Presence of parks in the zone	–	0.312 (2.94)
Correlation between weekday and weekend	0.394 (9.53)	
Spatial autoregressive parameter (δ)	0.561 (29.74)	
Composite marginal log-likelihood	−250,389.50	

Note: – = corresponding coefficient was not statistically significantly different from zero at the 90% level of confidence.

of service. Another possible explanation could be that higher-income individuals conduct more social and recreational activities during the week (compared with lower income segments) and use ridesourcing to access these activities. Finally, as expected higher rates of vehicle ownership are associated with a decrease in the generation of ridesourcing trips, a result also observed by Dias et al. (*4*) and Lavieri et al. (*5*).

The second half of Table 2 presents the threshold results. The elements of the $\boldsymbol{\alpha}$ vector do not have any substantive interpretations, but play the very important role of accommodating high- or low-probability masses for specific outcomes The elements in the $\boldsymbol{\gamma}$ vector are presented next in Table 2. The constants within the $\boldsymbol{\gamma}$ vector do not have any particular interpretation. For the other variables, a positive coefficient shifts all the thresholds toward the left of the demand intensity scale [see Castro et al. (*10*) for a detailed discussion], which has the effect of reducing the probability of the zero-trip count. A negative coefficient, on the other hand, shifts all thresholds toward the right of the generation propensity scale, which has the effect of increasing the probability of the zero count. We observe that the proportion of the male population in a zone has opposite effects on weekdays and weekends. Zones that have a higher male population proportion are more likely (than zones with a higher female population proportion) to have non-zero ridesourcing trips during the weekday and zero ridesourcing trips during the weekend days. In addition, zones with high vehicle ownership rates are more likely to have zero ridesourcing trips generated in a weekday, whereas zones that have parks are less likely to have zero trips generated in a weekend. Both results are expected, as having vehicles available in the household reduces the necessity of seeking alternative modes, whereas parks are associated with recreational activities that are more prevalent on weekends.

Table 3. Estimation Results of the Trip Distribution Split Model

Variables	Weekday	Weekend day
	Estimate (*t*-stat)	Estimate (*t*-stat)
(Constant)	−5.051 (−13.60)	−2.098 (−14.74)
Special land-use		
If all or part of the zone is occupied by UT	0.711 (2.85)	–
If the area is in the CBD	–	−0.673 (−4.37)
If the area contains the airport	4.142 (14.81)	–
If the area is outside the area of interest	3.119 (7.85)	–
Residential density		
Population density (natural logarithm of person per km^2)	0.110 (2.35)	–
Activity intensity		
Retail employment density (natural logarithm of retail jobs per km^2)	0.179 (3.33)	0.164 (3.34)
Population characteristics		
Proportion of White population	–	−1.653 (−4.90)
Median annual household income (divided by $10,000)	–	0.069 (3.72)
Proportion of households with two or more automobiles	−0.851 (−2.36)	−1.723 (−5.03)
Proportion of males	–	2.089 (3.94)
Trip characteristics		
Log of average cost between zones	−0.422 (−2.97)	−0.378 (−4.99)
Other characteristics		
Area (km^2)	0.367 (3.68)	0.218 (1.94)

Note: – = the corresponding coefficient was not statistically significantly different from zero at the 90% level of confidence.

Finally, at the bottom of the table we present the cross-correlation between weekdays and weekend days as well as the spatial autoregressive parameter. For the spatial correlation between zones we tested two different weight matrices, one based on the inverse of the distance between the centroid of the zones and another based on the inverse of the squared distance. The best model fit was obtained with the first weight matrix. The results confirm ours hypothesis that the number of trips that a zone generates in a weekday is positively associated with the number of trips generated on a weekend day. In addition, the number of trips in a zone is influenced by observed and unobserved factors of the neighboring zones.

Trip Distribution by Day Type

The University of Texas has a positive effect on trip attractions during the weekday, suggesting that people might be using ridesourcing to access the campus area at significantly higher rates than other zones. This effect seems to disappear, however, on weekends. This is very likely due to the reduced number of activities on campus during weekends, which results in less people visiting (and traveling to) the area. Whereas zones located in the CBD do not attract ridesourcing trips more so than zones in other areas of town, the demand to such CBD zones does decrease on weekends, likely due to the lack of activities during that period. The coefficients related to the airport and external zones are somewhat difficult to interpret directly because zones that fell into these categories had no associated data besides the trip cost. Therefore, many effects are entangled and cannot be interpreted immediately.

The proportion of retail employment, regardless of the day of the week, has a positive impact on trip attractions. Retail employment may be viewed as a proxy for opportunities for out-of-home activities: people will tend to travel more to places that have activities they want to partake in, such as shopping, and this appears to have a direct positive association with ridesourcing destination points. As expected, there is a positive influence of population density on trip attraction, representing return-home trips. Curiously, though, this effect is not statistically significant during the weekend. This could be a simple reflection of the higher number of out-of-home activities pursued during the weekend days. Thus, even if there are more ridesourcing return-home trips on weekend days than on weekdays, the proportion of such trips (as a fraction of total trips) may be lower on weekend days.

As explained previously, the negative effect of the proportion of the white population in a zone on ridesourcing trips generated from the zone may be a reflection of an intrinsic dislike for non-private travel (that is, a generic private auto-inclination). Similarly, the results in Table 3 indicate that, as the proportion of households with two or more vehicles in a zone increases, the "attractiveness" of the zone as a terminating point for ridesourcing trips decreases.

The average monetary cost of the trip plays a significant role in the trip distribution process for both weekends and weekdays. Throughout the estimation process, both distance and cost were used, but these two variables were too strongly correlated for both of them to be statistically significant. Therefore, given that the cost variable successfully explained most of the variance of these two variables, the distance variable was omitted from the estimation. The final variable is a pure size effect.

Conclusions

This paper has undertaken an analysis of the demand for ridesourcing trips in the city of Austin, Texas. Based on data provided by a non-profit TNC that entered Austin's market after the withdrawal of Uber and Lyft, we develop two models that analyze characteristics of the generation and distribution of ridesourcing trips at a TAZ level. Several public datasets were compiled to complete the analysis, including TAZ-level demographic data obtained from the Capital Area Metropolitan Planning Organization, the most recent Census estimates, and GTFS available from the state of Texas website. The use of open-source data is in its early stages and this paper provides a first glimpse of the potential that these data sources have in informing transportation models.

Our model provides important initial insights on characteristics of ridesourcing demand. In addition, it identifies interesting heterogeneities between ridesourcing use on weekdays and weekend days. For example, in the context of university campuses, our results suggest that students may be the beneficiaries of the availability of ridesourcing services. This may be either because vehicle ownership rates among university students is lower (compared with working individuals), or because of restricted parking regulations and high parking fees in such areas. Moreover, bus frequencies seem to have a negative impact on the generation of ridesourcing trips during the week, suggesting a substitution effect between ridesourcing and transit use. Another interesting finding is that the effect of the median household income in a zone on trip generation is opposite for weekdays and weekend days, suggesting that different income segments in the population may use ridesourcing for different activity purposes. Overall, the estimated parameters of the multivariate count model can be used to forecast the number of new ridesourcing trips in a TAZ in response to changing TAZ economics and demographics. The trip distribution model indicated that, as expected, the airport is a major ridesourcing trip attraction. This result leads to the question of whether only taxi and carpooling trips to the airport are being substituted, or if travelers who used to park at airports earlier are now opting for ridesourcing instead. A better understanding of this issue can help future parking planning at airports. Finally, the trip distribution model also provides evidence of the substantial use of weekday ridesourcing for returning home, which may suggest that ridesourcing is becoming integrated into the multi-modal use routine of individuals or is being used to avoid driving while impaired.

The results and methods used in this study can serve multiple purposes. First, from a travel behavior researcher perspective, this study has identified aggregate-level variables that have an impact on ridesourcing, and can guide efforts to better understand the demand for autonomous and connected vehicles in the future. Second, from a planner's perspective, this study has provided an analytic framework to develop predictive models of ridesourcing movements that can be accommodated in regional and planning network models. Finally, the results presented here may also have relevance to operators in their understanding of travel demand, which can lead to better strategies to allocate drivers to rides or to estimate optimal fleet size when entering a new market.

Acknowledgments

The authors would like to thank RideAustin for sharing the data publicly. This research was partially supported by the U.S. Department of Transportation through the Data-Supported Transportation Operations and Planning (D-STOP) Tier 1 University Transportation Center. The first author acknowledges funding support from CAPES and the Brazilian Government, and the fifth author acknowledges support from a Humboldt Research Award from the Alexander von Humboldt Foundation, Germany. The authors are grateful to Lisa Macias for her help in formatting this document, and to three anonymous referees who provided useful comments on an earlier version of the paper.

References

1. Uber Blog. Uber Hits 5 Billion Rides Milestone. www.uber.com/en-SG/blog/uber-hits-5-billion-rides-milestone/. Accessed July 26, 2017.
2. Rayle, L., D. Dai, N. Chan, R. Cervero, and S. Shaheen. Just a Better Taxi? A Survey-Based Comparison of Taxis, Transit, and Ridesourcing Services in San Francisco. *Transport Policy*, Vol. 45, 2016, pp. 168-178. https://doi.org/10.1016/j.tranpol.2015.10.004.
3. Clewlow, R. and G.S. Mishra. Shared Mobility: Current Adoption, Use, and Potential Impacts on Travel Behavior. Presented at the 96th Annual Meeting of the Transportation Research Board, Washington, D.C., 2017.
4. Dias, F.F., P.S. Lavieri, V.M. Garikapati, S. Astroza, R.M. Pendyala, and C.R. Bhat. A Behavioral Choice Model of the Use of Car-Sharing and Ride-Sourcing Services. *Transportation*, Vol. 44, No. 6, 2017, pp. 1307-1323. https://doi.org/10.1007/s11116-017-9797-8.
5. Lavieri, P.S., V.M. Garikapati, C.R. Bhat, R.M. Pendyala, S. Astroza, and F.F. Dias. Modeling Individual Preferences for Ownership and Sharing of Autonomous Vehicle Technologies. *Transportation Research Record: Journal of the Transportation Research Board*, 2017. 2665: 1–10.

6. RideAustin. Data File and Code Book. Available from: https://data.world/ride-austin/ride-austin-june-6-april-13. Accessed July 26, 2017.
7. Texas Government. Schedule Data for All Fixed Route MetroBus Lines and MetroRail. https://data.texas.gov/Transportation/GTFS/r4v4-vz24. Accessed July 26, 2017.
8. Hampshire, R.C., C. Simek, T. Fabusuyi, X. Di, and X. Chen.Measuring the Impact of an Unanticipated Suspension of Ride-sourcing in Austin, Texas. https://ssrn.com/abstract=2977969 or http://dx.doi.org/10.2139/ssrn.2977969. Accessed July 26, 2017.
9. Bhat, C.R., R. Paleti, and P. Singh. A Spatial Multivariate Count Model for Firm Location Decisions. *Journal of Regional Science*, Vol. 54, No. 3, 2014, pp. 462-502. https://doi.org/10.1111/jors.12101.
10. Castro, M., R. Paleti, and C.R. Bhat. A Latent Variable Representation of Count Data Models to Accommodate Spatial and Temporal Dependence: Application to Predicting Crash Frequency at Intersections. *Transportation Research Part B*, Vol. 46, No. 1, 2012, pp. 253-272. https://doi.org/10.1016/j.trb.2011.09.007.
11. Sivakumar, A., and C. Bhat. Fractional Split-Distribution Model for Statewide Commodity-Flow Analysis. *Transportation Research Record: Journal of the Transportation Research Board*, 2002. 1790: 80–88.
12. Gourieroux, C., A. Monfort, and A. Trognon. Pseudo Maximum Likelihood Methods: Theory. *Econometrica*, Vol. 52, No. 3, 1984, pp. 681-700.
13. Giuliano, G. Travel, Location and Race/Ethnicity. *Transportation Research Part A*, Vol. 37, No. 4, 2003, pp. 351-372. https://doi.org/10.1016/S0965-8564(02)00020-4.
14. Smart, M.J. A Nationwide Look at the Immigrant Neighborhood Effect on Travel Mode Choice. *Transportation*, Vol. 42, No. 1, 2015, pp. 189-209. https://doi.org/10.1007/s11116-014-9543-4.

The Standing Committee on Transportation Planning Applications (ADB50) peer-reviewed this paper (18-06531).

Article

TRR
JOURNAL OF THE TRANSPORTATION RESEARCH BOARD

Transportation Research Record
2018, Vol. 2672(46) 41–54

DOI: 10.1177/0361198118801349
journals.sagepub.com/home/trr
SAGE

Street Intersection Characteristics and Their Impacts on Perceived Bicycling Safety

Kailai Wang[1] **and Gulsah Akar**[1,2]

Abstract

Safety concerns are among the main issues that deter people from bicycling in the U.S. Earlier studies have explored the associations between characteristics of intersection design and bicyclists' safety perceptions. Research shows that there are significant links between bicycling choice, safety perceptions, bicycling experience levels, and socio-demographics. Yet the existing bicycling safety-rating models do not control for individuals' socio-demographics and their levels of bicycling experience, which are known to affect bicycling choice. This study develops a Perceived Bicycling Intersection Safety (PBIS) model which helps engineers, planners, and decision makers to understand better how a wide range of intersection features contribute to bicyclists' safety perceptions, controlling for socio-demographics and bicycling experience. The empirical analysis is based on an online visual survey conducted at the main campus of The Ohio State University through March and April 2017. Results showed that visual surveys are effective in capturing information about bicycling preferences. The paper concludes with recommendations for infrastructure decisions and suggestions for future research. The results of this study can help planners design street intersections that bicyclists will prefer. Our model can be applied elsewhere to test the effects of different intersection and street features.

Transport policies dominated by motor vehicle transport have led to a number of issues related to environment, energy, and sustainability over the past decades. Decision makers, transportation planners, and practitioners have been seeking ways to mitigate these adverse effects by encouraging alternative modes of travel. Encouraging the use of bicycles is increasingly regarded as an effective strategy to benefit living environments, reduce fuel consumption, and promote public health. Bicycling is financially affordable, physically possible by most people, and also a fast option for short distance trips. The perception of bicycling as an unsafe transport mode is a significant obstacle for promoting bicycling in the U.S., however. Perception of safety is one of the major factors that affect bicycling choice. Safety concern about on-road motor vehicle traffic is the major deterrent to bicycle commuting (*1–9*). Off-road cycleways and protected on-road cycling facilities can increase both perceived and actual bicycling safety (*10, 11*). Ma, Dill, and Mohr (*12*) found that the perception of the environment had a direct and significant effect on bicycling choice; however, the objective environment may only indirectly affect bicycling choice by influencing perceptions.

Deaths in bicycle crashes made up 2% of all traffic fatalities in the U.S. in 2012 (*13*), however, only 1% of all trips were bicycle trips (*14*). Furthermore, 30% of all cyclist fatalities occurred at intersections (*13*). One plausible interpretation of this result is the high level of interactions between traffic flows and bicycles at intersections. Intersections are the most complex features in an urban transportation system (*15–17*). A critical question arises: how do intersection design characteristics affect bicyclists' perceptions of hazard or safety?

Earlier studies have identified the associations between detailed intersection design characteristics and bicyclists' safety perceptions. To date, the most comprehensive publication is the Pedestrian and Bicycle Intersection Safety Index Report of 2006 (*18*). The model used investigates how intersection attributes, such as traffic volumes, number of lanes, speed limits, and presence of bicycle lanes, parking, and traffic control devices contribute to cyclists' perceptions of safety at intersections. Another important indicator, bicycle level of service (BLOS), is also widely

[1]City and Regional Planning, Knowlton School, The Ohio State University, Columbus, OH
[2]Civil, Environmental and Geodetic Engineering, The Ohio State University, Columbus, OH

Corresponding Author:
Address correspondence to Kailai Wang: wang.7684@osu.edu

used to measure the effects of multiple design and operating features of urban roadways (e.g., width, motor vehicle volumes, and speeds) on the quality of bicycle service (*19*). Prior studies, however, failed to control for individuals' socio-demographic characteristics and levels of cycling experience, which are known to affect bicycling decisions. The literature suggests that the perceived safety of an experienced cyclist may differ from that of a novice cyclist. There also exist gender differences in bicycling risk perceptions (*1*, *20*). Infrastructure investments may be wasted and the effectiveness of policy interventions may be reduced if policy makers view the influences of cycling-supportive treatments as the same to all people.

Our study contributes to the literature by quantifying the impact of each intersection feature to safety perceptions of bicycling, controlling for socio-demographics and bicycling experience levels. We conducted an online visual survey on the main campus of The Ohio State University. Visual surveys are effective substitutes for field surveys which involve participants riding through the study sites (*21*, *22*). We collected information from a large group of individuals on their perceptions of safety at various intersections. We also collected data on other factors that are known to affect bicycling decisions, such as socio-demographic characteristics and bicycling experience levels. We expect that the research results can be used by transport engineers and policy makers to project the changes in people's perceptions of bicycling safety with respect to changes in intersection features. The research results could also provide recommended intersection scenarios for decision makers and practitioners to establish guidelines and standards regarding perceptions of bicycling intersection safety.

Literature Review

The connection between street intersection design and bicycling safety lies at the center of efforts to improve bicycle travel environments (*18*, *23*). Several mathematical models have been developed to study bicycling safety. Most of these models examine the determinants of level of service (LOS) or comfort level for bicycling, focusing on whether the bicycling activities are compatible with physical environments (*21*, *24–29*). A few studies have analyzed the links between intersection characteristics and the perceived safety of cyclists (*16*, *18*, *23*, *30*). These studies focus on developing models by quantifying cyclists' perceptions of hazard or safety at intersections. These studies are discussed below.

Landis et al. (*23*) developed the first statistically calibrated BLOS model for roadway segments based on real-time perceptions from 145 cyclists nationwide. Cyclists' comfort and safety ratings were scaled from A to F. In that model, the number of bicycle lanes, the number of traffic lanes and on-street parking space, the presence of a stripe separating the motor vehicles and bicycles, and pavement surface were found to be significant predictors of bicycling safety perceptions.

Harkey et al. (*30*) developed a bicycle compatibility index (BCI) for urban and rural roadways at midblock locations. The sites selected for the study are located in five cities that represent a range of geographic conditions in the U.S. Participants were asked to watch videos of various roadway segments and provide ratings of how comfortable they would feel riding on each segment. BCI model covers some additional factors that may affect cyclists' perceived levels of comfort and safety, such as curb lane width, traffic speed, and type of roadside development. Harkey et al. (*30*) transformed the estimated BCI values into bicycle LOS classifications. LOS A indicates that a roadway is extremely comfortable for an average cyclist. This transformation is based on the LOS criteria for vehicles (*19*), but focuses on bicycling activities.

Landis et al. (*16*) proposed the intersection LOS model for bicycle through movement. This is the first model that focused on the complex intersection features with a viewpoint of the whole transportation system. This model provides insight on intersection design characteristics that could more safely accommodate cyclists. Data were collected from cyclists who rode through 18 selected signalized intersections and recorded their comfort and safety ratings on a scale of A through F. The roadway traffic volume, the total width of the outside through lane, and the intersection crossing distance were found to be the primary factors affecting the bicycling safety perceptions at the intersection level. It is of interest to note that the presence of a bicycle lane or a paved shoulder was not found to be statistically significant.

The Bicyclist Intersection Safety Index (Bike ISI) developed by Carter et al. (*18*) is a more comprehensive safety-rating model for intersections compared with the other models. This model involves both subjective user ratings and objective data, such as evasive actions that are taken by cyclists to avoid a collision. In that study, 67 intersection approaches in four cities were selected, and the study received 97 safety ratings with the six-point scale from bicycling experts. The Bike ISI models were estimated for three possible bicycle movements at intersections: through movement, right turn, and left turn. Traffic volumes, number of lanes, speed limits, and presence of bicycle lanes, parking, and traffic control devices were found to affect Bike ISI values.

In general, the number of studies focusing on perceived bicycling intersection safety is limited and the models discussed above have similar limitations. First, these models do not cover several common features of street intersections in the U.S. which are known to affect bicycling patterns, such as the presence of roundabouts, bicycling crossings, bicycle boxes, and traffic diverters. Bicycle boxes and traffic diverters are two common types of intersection features in North American cities, but their effects on bicycling safety have not been assessed by empirical studies. Some studies promote roundabouts as features for increasing bicycling safety. To investigate the effects of roundabouts on bicycling safety, some researchers conducted before-and-after studies in Europe. Schoon and Minnen (*31*) report that, after converting intersections with traditional stop signs to roundabouts, there was an 8% reduction in cyclist crash rate together with a 30% reduction in injury rate. Wegman et al. (*32*) propose that converting three-leg or four-leg intersections to roundabouts would reduce the number of potential conflict points. A well-designed modern roundabout could mitigate the accident severity level because of narrower lane widths and slower travel speeds (*32*). Bicycle facilities (bicycle tracks/lanes/paths) also play an important role in promoting bicycling safety at or near roundabouts. In a study conducted in Flanders, Belgium, Daniels et al. (*33*) found that, for roundabouts with mixed traffic, bicycle tracks or grade-separated paths could increase bicycling safety while bicycle lanes may lead to higher collision risks. Reynolds et al. (*34*) state that the safest design strategy was to build roundabouts with separate bicycle tracks. Second, the models developed by previous studies do not control for socio-demographic characteristics and bicycling skills. The significant effects of these factors have been widely discussed in the literature (e.g., *1*, *20*, *35–37*). Baker (*35*) proposes that women are the "indicator species" for bicycle-friendly environments due to their risk perceptions. Krizek, Johnson, and Tilahun (*36*) examine how bicycling patterns differ between men and women. They find that overall women bicycle less than men. Their study also shows that women are more likely to bicycle for shopping trips, running errands, or visiting friends. Akar and Clifton (*1*) found that safety and travel time are two major factors affecting bicycling choices. Their study suggests that women are less likely to use bicycles for commuting. Emond, Tang, and Handy (*37*) propose that perceptions of safety and household responsibilities make women less likely to bicycle than men. Akar, Fisher, and Namgung (*20*) found different risk perceptions between men and women in similar environments and that women are more likely to cite "lack of bicycle infrastructure" as a deterrent to bicycling. They state that gender is highly correlated with bicycling experience levels (based on self-reports of the survey respondents) and women are more likely to self-identify as beginner cyclists. These characteristics have implications for the resulting bicycling choices; therefore, they need to be controlled for while evaluating the effects of other factors on bicycling decisions. The PBIS model developed in this study integrates these characteristics into the analysis. The Example Application section presents how our model can estimate the perceived safety scores for different individuals with varying socio-demographics and bicycling experience levels.

Data and Methods

The development of the PBIS model follows the steps listed below. These steps are described in detail in the subsequent sections:

- Select intersection images that cover a wide range of features.
- Conduct an online survey to capture individuals' perceived bicycling safety.
- Develop models to understand the determinants of PBIS scores.

Selection of Intersection Images

Each street intersection has different characteristics that affect cyclists' safety perceptions. Rather than providing a rating for the whole intersection, we aim to identify the effect of each intersection feature on the overall safety perception. Based on the existing literature, we classify the factors of interest into the following groups:

1. Types of intersections and traffic controls
2. Road traffic
3. Intersection treatments
4. Presence of cycle tracks
5. Presence of traffic diverters
6. Sidewalk environments
7. Characteristics of the surrounding environment
8. Green space

Intersection images used in this study are taken from Google Street View. We selected 90 intersection images from the metropolitan area of Columbus, Ohio, as well as other cities around the world. Each image presents a combination of several intersection-level attributes. Our final selected intersection images represented a variety of intersection designs and traffic conditions for comparative analysis. Figure 1 presents two examples. The upper image shows a dense roundabout with traffic signals. This image also illustrates the presence of a bike box, a crosswalk, sidewalks on both sides of the streets, as well as two-way cycle tracks. The lower image is a dense four-way signalized intersection with limited off-road facilities. Table 1 presents the street intersection features that are of interest to this study.

Figure 1. Examples of selected intersection images.

Visual Survey

We designed and conducted an online visual survey using Qualtrics, a web-based survey tool. Our survey was conducted on the main campus of The Ohio State University (OSU). We sent invitation emails to 20,000 individuals in late March (OSU students, faculty, and staff members). We followed up with a reminder email in mid April. The survey was open from March 22 to May 1 in 2017. Each participant was asked to rate a randomly selected set of twelve intersections among the total 90 intersection images on a five-point scale of "Very unsafe to cross" to "Very safe to cross." We received valid responses from 1,376 individuals. These respondents rated at least one of our intersection images and provided information on their socio-demographic characteristics, attitudes towards bicycling, and bicycling experience levels. A descriptive summary of the respondents' characteristics is presented in Table 2. The resulting sample represents a wide range of individuals—a range much broader than in previous studies.

Model Specification

In this study, the perception of bicycling intersection safety is defined as a typical ordinal variable that is scaled into five scores:

1. Very unsafe to cross
2. Unsafe to cross
3. Neutral
4. Safe to cross
5. Very safe to cross

Ordinal regression models have been widely used to fit data with ordinal responses (*38*). We employ an ordered probit (OP) model to investigate the

Table 1. Variables of Interest

Group	Variable and its description[a]	Mean or percentage across the 90 images[b]
Types of intersections and traffic signal controls	3-way intersection with traffic signals	4.4%
	3-way intersection without traffic signals	8.9%
	4-way intersection with traffic signals	56.7%
	4-way intersection without traffic signals	18.9%
	Complicated intersection (≥5 ways) with traffic signals	3.3%
	Complicated intersection (≥5 ways) without traffic signals	2.2%
	Roundabout with traffic signals	1.1%
	Roundabout without traffic signals	4.4%
Curbs, lanes, and road traffic	Width of curb lane	
	The same as the central lanes	37.5%
	Wider than central lanes	33.3%
	Hold 2 vehicles in a row	12.5%
	Hold more than 2 vehicles in a row	16.7%
	Number of through auto lanes	
	The number of through auto lanes is 1	23.3%
	The number of through auto lanes is 2	34.4%
	The number of through auto lanes is 3	26.7%
	The number of through auto lanes is 4	11.1%
	The number of through auto lanes is 5	4.4%
	Main traffic volume (Measures: number of moving vehicles on through auto lanes/number of through auto lanes)	0.77
	Cross traffic volume (Measures: number of moving vehicles on crossing lanes)	1.13
	One-way through lanes	17.8%
	One-way crossing lanes	16.7%
	Left turning lane	20.0%
	Right turning lane	5.6%
	Marked left turning crossing	10.0%
	Marked right turning crossing	5.6%
	Posted speed limit	3.3%
Intersection treatments	Stop sign	13.3%
	Number of marked bicycle crossings	
	The number of marked bicycle crossings is 0	86.7%
	The number of marked bicycle crossings is 1	11.1%
	The number of marked bicycle crossings is 2	2.2%
	Marked pedestrian crosswalk sign	18.9%
	Two-stage turn box	30.2%
	Bike box	36.5%
	Intersect crossing markings	41.8%
Intersection treatments	Through median refuge island	18.1%
	Crossing median refuge island	25.1%
	Crosswalk	80.0%
Presence of cycle tracks	Two-way cycle tracks	10.0%
	One-way cycle tracks	10.0%
	On-street bike lanes	23.3%
	Multi-use paths	8.9%
Presence of traffic diverters	Traffic-calming circle	6.7%
	Forced turning islands	2.2%
	Diagonal diverter	2.2%
Sidewalk environments	Two-side sidewalk	90.0%
	Curb ramp	46.7%
	Clear curb extension	40.0%
	Paved shoulder	20.7%

(continued)

Table 1. (continued)

Group	Variable and its description[a]	Mean or percentage across the 90 images[b]
Surroundings	On-street parking	47.8%
	Slope	12.2%
	Street lights	95.6%
	Visible highway crossing	2.2%
	Light rail passing	3.3%
	Under maintenance	2.2%
	Bus stops	4.4%
	Bicycle racks & bike sharing stations	8.9%
	Near gas stations	4.4%
Green space	No green space	13.3%
	Grass borders along the road	5.6%
	Grass borders along the road, as well as trees behind the sidewalk	11.1%
	Grass borders with trees along the road	12.2%
	Grass borders and trees along the road, as well as trees behind the sidewalk	2.2%
	Trees on the sidewalk	28.9%
	Trees behind the sidewalk	18.9%
	Trees on the sidewalk, as well as behind the sidewalk	2.2%
	Heavy set of trees	5.6%
Visible active traveler	Presence of bicyclists	17.8%
	More than 2 visible pedestrians	30.0%

[a]Variables without a description in parenthesis are dummy variables: 1, if present; 0, otherwise.
[b]For continuous variables, we report the mean values. We report the percentages for categorical variables. For instance, for the first characteristic "3-way intersection with traffic signals," the reported percentage is 4.4%. This means out of 90 images, four images present this feature (90 × 4.4%).

relationships between safety perceptions and our factors of interest. Following Washington et al. (38), the OP models are derived by an unobserved latent variable z, which is used as the basis for modeling the safety perception rankings. The latent variable z is defined as a linear function for each observation n:

$$z_n = \beta X_n + \varepsilon_n \tag{1}$$

where X_n is a vector of explanatory variables that determine the safety perception, β is a vector of coefficients, and ε_n is a random error term following the standard normal distribution. Based on this equation, the scores of bicycling safety perception Y for each observation are defined as:

$$Y = \begin{cases} 1 & if\ z_n \leq \tau_1 \\ 2 & if\ \tau_1 < z_n \leq \tau_2 \\ 3 & if\ \tau_2 < z_n \leq \tau_3 \\ 4 & if\ \tau_3 < z_n \leq \tau_4 \\ 5 & if\ z_n > \tau_4 \end{cases} \tag{2}$$

where τ is the threshold parameter (cut-off point) to be estimated for each score. The probability of dependent variable Y taking on each score of safety perception is:

$$\begin{aligned} P(Y = 1) &= \Phi(\tau_1 - \beta X_n) \\ P(Y = 2) &= \Phi(\tau_2 - \beta X_n) - \Phi(\tau_1 - \beta X_n) \\ P(Y = 3) &= \Phi(\tau_3 - \beta X_n) - \Phi(\tau_2 - \beta X_n) \\ P(Y = 4) &= \Phi(\tau_4 - \beta X_n) - \Phi(\tau_3 - \beta X_n) \\ P(Y = 5) &= 1 - \Phi(\tau_4 - \beta X_n) \end{aligned} \tag{3}$$

where $P(Y = j)$ is the probability of safety perception j, $j = 1, 2, 3, 4, 5$. $\Phi(.)$ is the cumulative distribution function for standard normal distribution. The coefficient of each variable and threshold parameters can be determined by the maximum likelihood estimation. Note that because higher scores indicate safer environments, the interpretation of an estimated coefficient is such that positive signs mean improved perceived bicycling safety scores. However, these coefficients cannot be intuitively interpreted as the effects of explanatory variables as in OLS (ordinary least squares) regression models. To present meaningful interpretations, we then compute the marginal effect of each variable to quantify the impacts on each category of the dependent variable Y. The marginal effect of a variable illustrates the change of probability of each safety level caused by one unit increase in the input variable, while keeping other variables at fixed values, as follows:

Table 2. Descriptive Summary of Respondents' Characteristics

Survey items	Number of responses	Answers	Percentage
Primary mode of transportation	1373	Auto (drive alone)	56.6%
		Carpool (with one or more people)	5.3%
		Bus	5.3%
		Walking	19.1%
		Bicycle	11.9%
		Other	1.8%
Last time that you rode a bicycle (for any purpose)	1375	Within last week	26.9%
		Within last month	9.8%
		Within the last year	30.5%
		More than a year ago	30.0%
		Never	2.8%
Bicycling experience levels	1211	I cannot ride a bicycle	2.5%
		A novice cyclist	34.6%
		An intermediate cyclist	35.2%
		An advanced, confident cyclist	23.5%
		I don't know how to describe my bicycling skills	4.2%
Age	1285	18–25 years old	35.6%
		26–35 years old	46.7%
		36–50 years old	3.3%
		51 years and older	14.4%
Gender	1298	Male	37.9%
		Female	62.1%
International student/staff/faculty	1256	International	11.0%
		Non-international	89.0%
Status at OSU	1233	Faculty	9.7%
		Staff	43.0%
		Graduate/Professional student	18.1%
		Undergraduate student	28.5%
		Others	0.7%

$$\partial P(Y=1)/\partial X = -\phi(\tau_1 - \beta X)\beta$$
$$\partial P(Y=2)/\partial X = [\phi(\tau_1 - \beta X) - \phi(\tau_2 - \beta X)]\beta$$
$$\partial P(Y=3)/\partial X = [\phi(\tau_2 - \beta X) - \phi(\tau_3 - \beta X)]\beta$$
$$\partial P(Y=4)/\partial X = [\phi(\tau_3 - \beta X) - \phi(\tau_4 - \beta X)]\beta$$
$$\partial P(Y=5)/\partial X = \phi(\tau_4 - \beta X)\beta \quad (4)$$

where $\phi(.)$ is the probability density function of the standard normal distribution.

Discussion of Model Results

The results of the OP model are presented in Table 3. Variables of interest that were not statistically significant at the 90% level or higher were omitted from the final PBIS model. Table 4 presents the average marginal effects of the explanatory variables.

Types of Intersections and Traffic Signal Controls

As expected, complicated intersections (more than four ways) without traffic signals are negatively associated with individuals' PBIS scores. However, riding through complicated intersections with traffic signals does not significantly affect safety perceptions as compared with three-way intersections with/without traffic signals and four-way intersections with/without traffic signals. This result is not surprising. Complicated intersections include more conflict points than three- or four-way intersections. Hence, people are less likely to feel safe while crossing. This effect may be counteracted by the presence of traffic signals. Our results suggest that individuals would feel safer riding through roundabouts compared with traditional intersections. We find that, all else being equal, riding through a roundabout with traffic signals instead of a traditional intersection increases the likelihood that one would find this scene "safe" by 8.1% and "very safe" by 7.1%.

Road Traffic

The number of through auto lanes is negatively associated with the perceived bicycling safety. This may be because more auto lanes expose cyclists to more moving vehicles. A high main traffic volume also has a negative effect on safety. On the other hand, crossing a one-way street is positively associated with perceived bicycling safety. This is reasonable because one-way streets are

Table 3. OP Model of Perceived Bicycling Intersection Safety

Variable	Coefficient	Robust std. err.	*P* value
Types of intersections and traffic signal controls[a]			
Complicated intersection without traffic signals	−0.301	0.078	0.000
Roundabout with traffic signals	0.430	0.141	0.002
Roundabout without traffic signals	0.544	0.105	0.000
Road traffic			
Number of through auto lanes	−0.203	0.032	0.000
Main traffic volume	−0.190	0.046	0.000
One-way crossing lanes	0.150	0.076	0.049
Intersection treatments			
Marked or unmarked crosswalk	−0.167	0.093	0.075
Number of marked bicycle crossings	0.281	0.167	0.093
Number of marked bicycle crossings × number of marked bicycle crossings	−0.161	0.087	0.066
Two-stage turn box	0.211	0.082	0.010
Bike box	0.359	0.064	0.000
Intersection crossing markings	0.229	0.071	0.001
Through median refuge island	−0.280	0.150	0.062
Crossing median refuge island	−0.802	0.091	0.000
Cycle tracks			
Two-way cycle tracks	0.307	0.120	0.010
One-way cycle tracks	0.267	0.097	0.006
Traffic diverters			
Traffic-calming circle	0.187	0.113	0.099
Forced turning islands	0.184	0.092	0.045
Sidewalk			
Presence of a two-side sidewalk	0.209	0.069	0.003
Curb ramp	0.150	0.066	0.022
Clear curb extension	0.192	0.063	0.002
Paved shoulder	0.623	0.107	0.000
Surroundings			
Light rail passing	−0.362	0.116	0.002
Visible highway crossing	−0.223	0.082	0.007
Green space[b]			
Grass borders with trees along the road	0.338	0.075	0.000
Grass borders along the road, as well as trees behind the sidewalk	0.333	0.111	0.003
Grass borders with trees along the road, as well as trees behind the sidewalk	1.016	0.059	0.000
Trees on the sidewalk, as well as behind the sidewalk	0.245	0.112	0.028
Heavy set of forest	0.438	0.123	0.000
Visible active travelers			
Presence of bicyclists	0.164	0.076	0.030
More than two visible pedestrians	−0.149	0.064	0.020
Age[c]			
26–35 years old	−0.108	0.026	0.000
36–50 years old	−0.139	0.029	0.000
51 years and older	−0.155	0.038	0.000
Gender			
Male	0.063	0.024	0.007
Bicycle experience level[d]			
Intermediate cyclist	0.241	0.024	0.000
Advanced, confident cyclist	0.450	0.032	0.000
Threshold_cut1	−1.646	0.119	
Threshold_cut2	−0.550	0.119	
Threshold_cut3	0.130	0.118	
Threshold_cut4	1.417	0.121	
Summary statistics			
Number of observations		12839	
Initial log likelihood		−19130.441	
Final log likelihood		−17550.965	
Cragg & Uhler's pseudo R2		0.202	

[a]Baseline: 3-way intersection with/without traffic signals; 4-way intersection with/without traffic signals; complicated intersection with traffic signals.
[b]Baseline: no green space; grass borders along the road; trees on the sidewalk; Trees behind the sidewalk.
[c]Baseline: 18–25 years old.
[d]Baseline: I cannot ride a bicycle; a novice cyclist; I don't know how to describe my bicycling skills.

Table 4. Average Marginal Effects for OP Model of PBIS

Variable	Marginal effect				
	Very unsafe	Unsafe	Neutral	Safe	Very safe
Types of intersections and traffic signal controls[a]					
Complicated intersection without traffic signals	3.6%	5.8%	1.2%	−5.6%	−5.0%
Roundabouts with traffic signals	−5.1%	−8.3%	−1.8%	8.1%	7.1%
Roundabouts without traffic signals	−6.4%	−10.5%	−2.2%	10.2%	9.0%
Road traffic					
Number of through auto lanes	2.4%	3.9%	0.8%	−3.8%	−3.4%
Main traffic volume	2.2%	3.7%	0.8%	−3.6%	−3.1%
One-way crossing lanes	−1.8%	−2.9%	−0.6%	2.8%	2.5%
Intersection treatments					
Marked or unmarked crosswalk	2.0%	3.2%	0.7%	−3.1%	−2.8%
Number of marked bicycle crossings	−3.3%	−5.4%	−1.1%	5.3%	4.6%
Number of marked bicycle crossings × number of marked bicycle crossings	1.9%	3.1%	0.7%	−3.0%	−2.7%
Two-stage turn box	−2.5%	−4.1%	−0.9%	3.9%	3.5%
Bike box	−4.2%	−6.9%	−1.5%	6.7%	5.9%
Intersection crossing markings	−2.7%	−4.4%	−0.9%	4.3%	3.8%
Through median refuge island	3.3%	5.4%	1.1%	−5.2%	−4.6%
Crossing median refuge island	9.5%	15.5%	3.3%	−15.0%	−13.3%
Cycle tracks					
Two-way cycle tracks	−3.6%	−5.9%	−1.3%	5.7%	5.1%
One-way cycle tracks	−3.1%	−5.2%	−1.1%	5.0%	4.4%
Traffic diverters					
Traffic-calming circle	−2.2%	−3.6%	−0.8%	3.5%	3.1%
Forced turn islands	−2.2%	−3.6%	−0.8%	3.4%	3.0%
Sidewalk environments					
Presence of a two-side sidewalk	−2.5%	−4.0%	−0.9%	3.9%	3.5%
Curb ramp	−1.8%	−2.9%	−0.6%	2.8%	2.5%
Clear curb extension	−2.3%	−3.7%	−0.8%	3.6%	3.2%
Paved shoulder	−7.3%	−12.0%	−2.5%	11.7%	10.3%
Surroundings					
Light rail passing	4.3%	7.0%	1.5%	−6.8%	−6.0%
Visible highway crossing	2.6%	4.3%	0.9%	−4.2%	−3.7%
Green space[b]					
Grass borders with trees along the road	−4.0%	−6.5%	−1.4%	6.3%	5.6%
Grass borders along the road, as well as trees behind the sidewalk	−3.9%	−6.4%	−1.4%	6.2%	5.5%
Green space[b]					
Grass borders with trees along the road, as well as trees behind the sidewalk	−12.0%	−19.7%	−4.1%	19.0%	16.8%
Trees on the sidewalk, as well as behind the sidewalk	−2.9%	−4.7%	−1.0%	4.6%	4.1%
Heavy set of forest	−5.2%	−8.5%	−1.8%	8.2%	7.2%
Visible active travelers					
Presence of bicyclists	−1.9%	−3.2%	−0.7%	3.1%	2.7%
More than two visible pedestrians	1.8%	2.9%	0.6%	−2.8%	−2.5%
Age[c]					
26–35 years old	1.3%	2.1%	0.4%	−2.0%	−1.8%
36–50 years old	1.6%	2.7%	0.6%	−2.6%	−2.3%
51 years old and older	1.8%	3.0%	0.6%	−2.9%	−2.6%
Gender					
Male	−0.7%	−1.2%	−0.3%	1.2%	1.0%
Bicycle experience level[d]					
Intermediate cyclist	−2.8%	−4.7%	−1.0%	4.5%	4.0%
Advanced, confident cyclist	−5.3%	−8.7%	−1.8%	8.4%	7.4%

[a]Baseline: 3-way intersection with/without traffic signals; 4-way intersection with/without traffic signals; complicated intersection with traffic signals.
[b]Baseline: no green space, grass borders along the road; trees on the sidewalk; trees behind the sidewalk.
[c]Baseline: 18–25 years old.
[d]Baseline: I cannot ride a bicycle; a novice cyclist; I don't know how to describe my bicycling skills.

designed to have the traffic flow in one direction. Bicyclists would only need to watch out for one directional traffic flow, which decreases the potential automobile-involved conflict points for turning.

Intersection Treatments

The number of marked bicycle crossings is positively associated with perceived safety levels. This is because intersections that have raised and coloured bicycle crossings may lead to better recognition and perception of safety compared with other intersections. However, the negative relationship between the quadratic term of the number of marked bicycle crossings and respondents' safety perceptions suggests that installing several bicycle crossings at an intersection may make people feel less safe about bicycling there. One possible explanation for this result is that too many crossings may be confusing to both motorists and bicyclists (*39*). Both the presence of bicycle boxes and two-stage turn boxes at an intersection can significantly increase people's safety perceptions of bicycling there. Furthermore, the results of marginal effects reveal that people may feel safer when they are bicycling through an intersection with bicycle boxes than with a two-stage turn box. Holding everything else constant, installing a two-stage turn box increases the probability that one would find this intersection "safe" by 3.9% and "very safe" by 3.5%. With respect to installing a bicycle box, the likelihood of reporting "safe" and "very safe" will increase by 6.7% and 5.9%, respectively. This is probably because bicycle boxes offer larger waiting areas for bicyclists, which improves their perceived safety. However, the cost of installing a bicycle box is typically much higher than that of a two-stage turn box (*40*). Future research should delve more deeply into the efficiency and effectiveness of these bicycle-friendly infrastructure investments.

The presence of intersection crossing markings may significantly improve the perceptions of bicycling safety. This is not surprising because intersection crossing markings indicate the intended path of a bicyclist, which may guide them through the intersection in a straight path. This treatment contributes to reducing stress for bicyclists at intersections by delineating the bicycling zone (*41*). Interestingly, our results suggest that both the presence of a through median refuge island and a crossing median refuge island at an intersection may have a negative effect on the perceived safety for bicycling. In our selected images, most of the median refuge islands are placed on well-marked crosswalks. Hence, the potential pedestrian traffic on these islands may force bicyclists to adjust their speed appropriately, which may affect their perceived safety. We find that the presence of crosswalks is negatively associated with perceived safety for bicycling.

Cycle Tracks

The presence of cycle tracks increased the respondents' perceptions of safety at intersections. A comparison of marginal effects between one-way cycle tracks ("safe" 5.0%; "very safe" 4.4%) and two-way cycle tracks ("safe" 5.7%; "very safe" 5.1%) indicates that converting one-way cycle tracks to two-way cycle tracks may slightly increase perceived safety for bicycling. This result might be because two-way cycle tracks allow bicycle movement in both directions, which may decrease the likelihood of bicycle collisions. Also, two-way cycle tracks always have bicycle traffic signals at intersections.

Traffic Diverters, Sidewalk Environments, Surroundings

The presence of a traffic-calming circle increases the perceived safety for bicycling at the intersection level. This is probably because this facility requires moving vehicles to turn and reduce speed instead of traveling straight ahead through an intersection. Hence, people may feel that the probability of collisions with automobiles would be reduced. Similarly, the presence of forced turn islands also promotes safety perceptions. With respect to the sidewalk environments, the presence of two-side sidewalks and curb ramps are positively associated with perceived safety for cycling. Compared with lack of curbs and unclear curb extensions, clear curb extensions and the presence of paved shoulders are associated with higher safety perceptions. For surrounding environments, we find that the presence of light rail passing and visible highway crossing are negative predictors of individuals' PBIS scores.

Green Space

The relationship of green space allocations and perceived safety for bicycling exhibits an interesting pattern. During the model development process, we tested the effects of grass borders along the road, trees on the sidewalk, and trees behind the sidewalk. We found no significant associations with these and safety perceptions. These factors were omitted from the final model and hence they became a part of the base case in our final reported PBIS model. Our final model suggests that grass borders with trees along the road, grass borders along the road, as well as trees behind the sidewalk, trees both on and behind the sidewalk, and heavy set of trees (forest) are positively associated with perceived safety for cycling at the intersection level as compared with our base cases. Furthermore, our results imply that grass borders with trees along the road, as well as trees behind the sidewalk (as in the lower image in Figure 1) have the highest impacts on individuals' safety perceptions.

Socio-Demographics and Bicycling Experiences

Our model controls for the effects of socio-demographics and bicycling experience levels which are known to affect

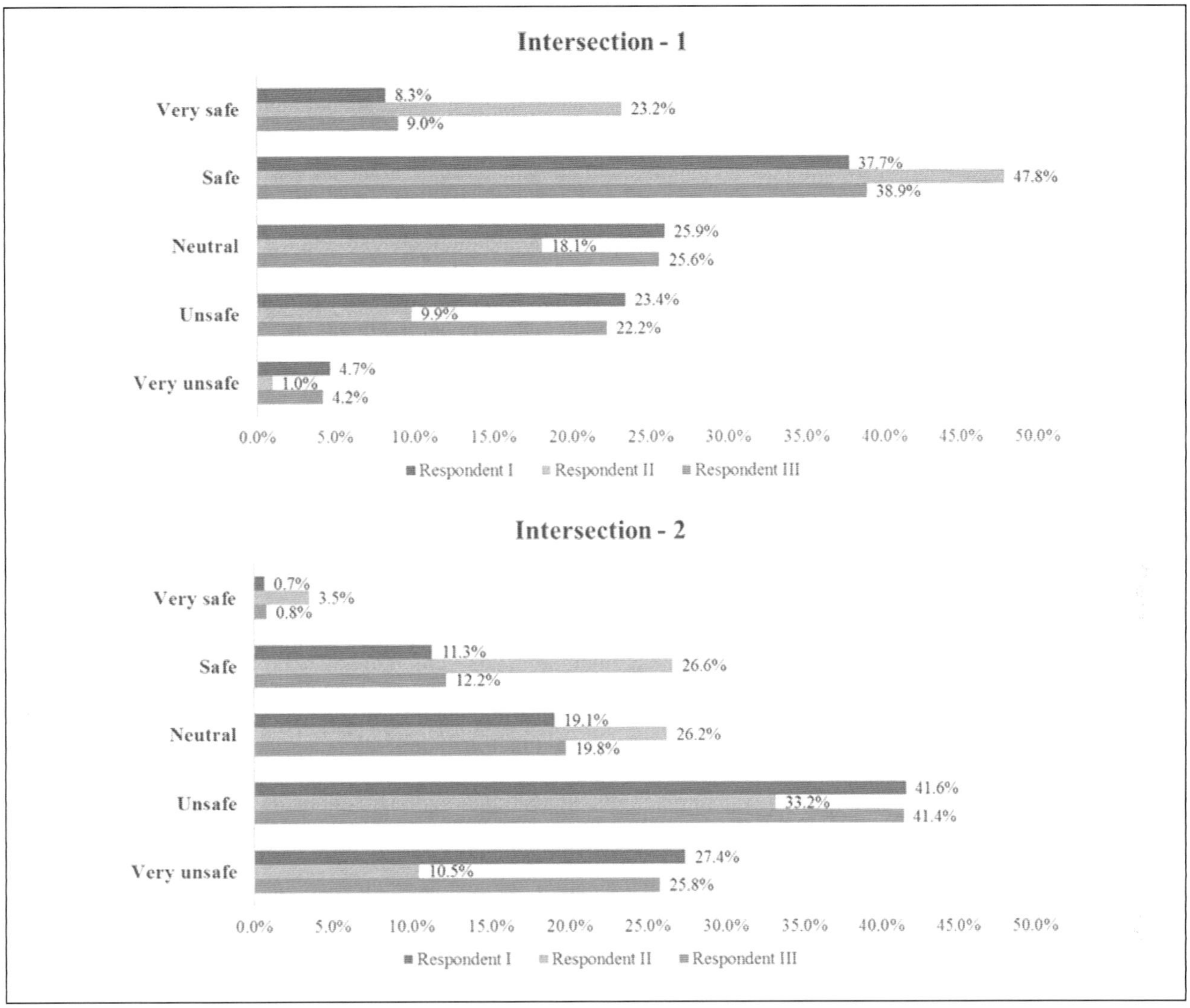

Figure 2. Probabilities of each level of safety perceptions for three respondents when riding through two intersections.

bicycling choices. Our results show that, all else being equal, younger individuals (18 to 25 years old) are more likely to feel safe than other age groups. As expected, men rate the same intersection images with higher safety scores than women. Our results also indicate that, all else being equal, individuals who identify themselves as intermediate or advanced cyclists are more likely to give higher safety scores than novice cyclists.

Example Application

As discussed, this study contributes to the literature by developing a bicycling intersection safety index accounting for people's socio-demographics and bicycling experience levels. To illustrate how our model works, we calculate the perceived safety scores for two images (as shown in Figure 1) based on three hypothetical individuals. Our hypothetical individuals represent different ages, genders, and bicycling experience levels. The descriptions are as follows:

- Individual I is a 62-year old female. She "doesn't know how to describe" her bicycling skills.
- Individual II is a 22-year old male. He describes himself as an advanced, confident cyclist who is comfortable riding in most traffic situations.
- Individual III is a 37-year old female. She describes herself as an intermediate cyclist who is somewhat comfortable riding in some traffic situations.

We apply our model for each individual. Figure 2 illustrates the probabilities of how safe each respondent may feel when riding through these two intersections. Our results show clear differences across these individuals and intersections. All individuals would find Intersection 1 safer than Intersection 2. As compared with other individuals, Individual II (young male, describes himself as an advanced cyclist) is more likely to have higher levels of perceived safety when riding through both intersections. This result highlights the necessity of controlling for socio-demographics and bicycling experience to determine the true and varying effects of intersection features on safety perceptions.

Policy Implications

Most of the factors contributing to our PBIS model correspond with prior studies and technical reports. For example, provision of transportation infrastructure—marked bicycling crossings, two-stage turn boxes, bike boxes, intersection crossing markings, cycle tracks, traffic diverters, two-side sidewalks, curb ramps, and paved shoulders—will improve people's perceptions of bicycling safety at the intersection level. While this result may be common sense to practitioners and bicyclists, a scientific verification with a concrete statistical approach can provide more detailed design guidelines for bicycle-friendly infrastructure investments.

One notable finding is that both through and crossing median refuge islands have negative impacts on perceived safety for bicycling. A median refuge island provides protected space for pedestrians and bicyclists to wait for an acceptable gap in traffic, which may allow them to cross streets more comfortably. However, the potential pedestrian volumes may make people feel less safe about bicycling. Our model further suggests that converting traditional three- or four-way intersections into roundabouts (when possible) as well as installing traffic signals may increase the safety perceptions of cycling in them.

It is worth noting that planting grass borders and trees along the road, as well as trees behind the sidewalk, is found to have the highest impact on increasing the perceived bicycling intersection safety. To our knowledge, this is the first analysis that explores the relationship between green space allocations and bicycle safety perceptions at the intersection level. This finding may be of great assistance to practitioners and policy makers for effectively allocating scarce resources, especially on investments for creating bicycle-friendly environments.

Our analysis suggests that the width of the curb lane, stop signs, and posted speed limits do not exhibit significant influences on bicycle safety perceptions at the intersection level. These results are inconsistent with some previous studies. For the width of the curb lane, the result may be due to the quality of our visual survey. Our intersection images are taken from different angles, and therefore it may be difficult for the survey respondents to figure out whether the curb lane is wider than other lanes. With respect to the presence of stop signs and posted speed limits, both items are highly correlated with other intersection features, hence they were dropped from our PBIS model.

In addition to the aforementioned general implications, this analysis investigates the contributions of socio-demographics and bicycling experiences to perceived safety of intersections. Our model aids in explaining how different combinations of intersection characteristics affect the resulting perceived bicycling safety scores for various groups. For example, all else being equal, females are less likely to give higher safety ratings. All else being equal, intermediate and advanced cyclists are more likely to give higher safety scores.

Conclusion

This article presents a comprehensive statistical model to predict how safe an individual is likely to feel about riding a bicycle in a given street intersection. Although this research is not the first to develop a model for bicycling safety index at intersections, it is the first model that accounts for individuals' socio-demographics and bicycling experience. Our model is an extension of other bicycling intersection safety models with a wider range of intersection features being considered. In addition, our study also controls for individuals' socio-demographics and bicycling experience.

The empirical analysis is based on the data collected via an online visual survey at the main campus of The Ohio State University through March and April 2017. An OP model is estimated to predict the probability that a user will provide a given safety score for an intersection. The estimated model covers a wider range of intersection features that are known to affect bicycling safety perceptions than previous studies. The developed model can be used to evaluate perceived bicycling safety for existing and planned street intersections. Decision makers can subsequently facilitate improvements to create bicycle-friendly intersections.

Future research related to quantifying bicycling safety at the intersection level should focus on exploring the relationship between safety perceptions and actual bicycle crashes. We would predict that intersections with low perceived bicycling intersection safety scores should be the ones with high accident rates.

Acknowledgments

This study was funded by a grant from the Ohio Department of Transportation (ODOT). The authors thank the ODOT associates and technical panel members for their time and insights throughout this study, and three anonymous reviewers for their useful comments. The authors also thank all survey respondents from The Ohio State University.

Author Contributions

The authors confirm contribution to the paper as follows: study conception and design: GA, KW; data collection: GA, KW; analysis and interpretation of results: KW, GA; draft manuscript preparation: KW, GA. Both authors reviewed the results and approved the final version of the manuscript.

References

1. Akar, G., and K. Clifton. Influence of Individual Perceptions and Bicycle Infrastructure on Decision to Bike. *Transportation Research Record: Journal of the Transportation Research Board*, 2009. 2140: 165–172.
2. Dill, J. Bicycling for Transportation and Health: The Role of Infrastructure. *Journal of Public Health Policy*, Vol. 30, 2009, pp. S95–S110.
3. Foster, C., J. Panter, and N. Wareham. Assessing the Impact of Road Traffic on Cycling for Leisure and Cycling to Work. *International Journal of Behavioral Nutrition and Physical Activity*, Vol. 8, No. 1, 2011, pp. 61–66.
4. Handy, S. L., M. G. Boarnet, R. Ewing, and R. E. Killingsworth. How the Built Environment Affects Physical Activity. Views from Urban Planning. *American Journal of Preventive Medicine*, Vol. 23, No. 2S, 2002, pp. 64–73.
5. Pucher, J., and R. Buehler. Making Cycling Irresistible: Lessons from the Netherlands, Denmark and Germany. *Transport Reviews*, Vol. 28, No. 4, 2008, pp. 495–528.
6. Pucher, J., and R. Buehler. Cycling Towards a More Sustainable Transport Future. *Transport Reviews*, Vol. 37, No. 6, 2017, pp. 689–694.
7. Sallis, J. F., T. L. Conway, L. I. Dillon, L. D. Frank, M. A. Adams, K. L. Cain, and B. E. Saelens. Environmental and Demographic Correlates of Bicycling. *Preventive Medicine*, Vol. 57, No. 5, 2013, pp. 456–460.
8. Schepers, P., M. Hagenzieker, R. Methorst, B. Van Wee, and F. Wegman. A Conceptual Framework for Road Safety and Mobility Applied to Cycling Safety. *Accident Analysis & Prevention*, Vol. 62, 2014, pp. 331–340.
9. Winters, M., M. Brauer, M. E. Setton, and K. Teschke. Built Environment Influences on Healthy Transportation Choices: Bicycling versus Driving. *Journal of Urban Health*, Vol. 87, No. 6, 2010, pp. 969–993.
10. Kerr, J., J. A. Emond, H. Badland, R. Reis, O. Sarmiento, J. Carlson, J. F. Sallis, E. Cerin, K. Cain, T. Conway, and G. Schofield. Perceived Neighborhood Environmental Attributes Associated with Walking and Cycling for Transport among Adult Residents of 17 Cities in 12 Countries: the IPEN Study. *Environment Health Perspect*, Vol. 123, No. 3, 2016, pp. 290–298.
11. Ma, L., and J. Dill. Associations between the Objective and Perceived Built Environment and Bicycling for Transportation. *Journal of Transport & Health*, Vol. 2, No. 2, 2015, pp. 248–255.
12. Ma, L., J. Dill, and C. Mohr. The Objective versus the Perceived Environment: What Matters for Bicycling? *Transportation*, Vol. 41, No. 6, 2014, pp. 1135–1152.
13. Traffic Safety Facts, 2013 Data. *Bicycling and Other Cyclists*. U.S. Department of Transportation, National Highway Traffic Safety Administration, Washington, D.C., 2014. https://crashstats.nhtsa.dot.gov/Api/Public/ViewPublication/812151.
14. National Household Travel Survey. *2009 NHTS*. U.S. Department of Transportation, Federal Highway Administration, Washington, D.C., 2011. http://nhts.ornl.gov/download.shtml.
15. Garber, N., and R. Srinivasan. Characteristics of Accidents Involving Elderly Drivers at Intersections. *Transportation Research Record: Journal of the Transportation Research Board*, 1991. 1325: 8–16.
16. Landis, B., V. Vattikuti, R. Ottenberg, T. Petritsch, M. Guttenplan, and L. Crider. Intersection Level of Service for the Bicycle through Movement. *Transportation Research Record: Journal of the Transportation Research Board*, 2003. 1828: 101–106.
17. Moore, D. N., W. Schneider, P. T. Savolainen, and M. Farzaneh. Mixed Logit Analysis of Bicycle Injury Severity Resulting from Motor Vehicle Crashes at Intersection and Non-Intersection Locations. *Accident Analysis and Prevention*, Vol. 43, No. 3, 2011, pp. 621–630.
18. Carter, D. L., W. W. Hunter, C. V. Zegeer, J. R. Stewart, and H. F. Huang. Cyclist Intersection Safety Index. *Transportation Research Record: Journal of the Transportation Research Board*, 2007. 2031: 18–24.
19. *Highway Capacity Manual*. Transportation Research Board of the National Academies, Washington, D.C., 2010.
20. Akar, G., N. Fischer, and M. Namgung. Bicycling Choice and Gender Case Study: The Ohio State University. *International Journal of Sustainable Transportation*, Vol. 7, No. 5, 2013, pp. 347–365.
21. Foster, N., C. M. Monsere, J. Dill, and K. Clifton. Level-of-Service Model for Protected Bike Lanes. *Transportation Research Record: Journal of the Transportation Research Board*, 2015. 2520: 90–99.
22. Zhu, W., B. Zhai, and D. Jian. Evaluating the Bicycle Travel Environment in a Changing Bicycle Culture: Case Study of Shanghai, China. *Journal of Urban Planning and Development*, Vol. 143, No. 3, 2017, p. 05017001. http://ascelibrary.org/doi/abs/10.1061/%28ASCE%29UP.1943-5444.0000377.
23. Landis, B., V. Vattikuti, and M. Brannick. Real-Time Human Perceptions Toward a Bicycle Level of Service. *Transportation Research Record: Journal of the Transportation Research Board*, 1997. 1578: 119–126.
24. Botma, H. Method to Determine Level of Service for Bicycle Paths and Pedestrian-Bicycle Paths. *Transportation Research Record: Journal of the Transportation Research Board*, 1995. 1502: 38–44.

25. Davis, J. *Bicycle Safety Evaluation*. Auburn University, City of Chattanooga, and Chattanooga-Hamilton County Regional Planning Commission, Chattanooga, TN, 1987.
26. Epperson, B. Evaluating Suitability of Roadways for Bicycle Use: Toward a Cycling Level-of-Service Standard. *Transportation Research Record: Journal of the Transportation Research Board*, 1994. 1438: 9–16.
27. Hunter, W. W., J. R. Stewart, and J. C. Stutts. Study of Bicycle Lanes versus Wide Curb Lanes. *Transportation Research Record: Journal of the Transportation Research Board*, 1999. 1674: 70–77.
28. Majumdar, B. B., and S. Mitra. Valuing Factors Influencing Bicycle Route Choice using a Stated-Preference Survey. *Journal of Urban Planning and Development*, Vol. 143, No. 3, 2017, p. 04017001.
29. Sorton, A., and T. Walsh. Bicycle Stress Level as a Tool to Evaluate Urban and Suburban Bicycle Compatibility. *Transportation Research Record: Journal of the Transportation Research Board*, 1994. 1438: 17–24.
30. Harkey, D., D. Reinfurt, and M. Knuiman. Development of the Bicycle Compatibility Index. *Transportation Research Record: Journal of the Transportation Research Board*, 1998. 1636: 13–20.
31. Schoon, C., and J. van Minnen. The Safety of Roundabouts in the Netherlands. *Traffic Engineering and Control*, Vol. 35, No. 3, 1994, pp. 142–148.
32. Wegman, F., F. Zhang, and A. Dijkstra. How to Make More Cycling Good for Road Safety? *Accident Analysis and Prevention*, Vol. 44, No. 1, 2012, pp. 19–29.
33. Daniels, S., T. Brijs, E. Nuyts, and G. Wets. Injury Crashes with Bicyclists at Roundabouts: Influence of Some Location Characteristics and the Design of Cycle Facilities. *Journal of Safety Research*, Vol. 40, No. 2, 2009, pp. 141–148.
34. Reynolds, C. C. O., M. A. Harris, K. Teschke, P. A. Cripton, and M. Winters. The Impact of Transportation Infrastructure on Bicycling Injuries and Crashes: A Review of the Literature. *Environmental Health*, Vol. 8, No. 47, 2009, pp. 1–19.
35. Baker, L. *How to Get More Bicyclists on the Road: To Boost Urban Cycling, Figure Out What Women Want*. Scientific American, Washington, D.C., 2009. https://www.scientificamerican.com/article/getting-more-bicyclists-on-the-road/.
36. Krizek, K. J., P. J. Johnson, and N. Tilahun. Gender Differences in Bicycling Behavior and Facility Preferences. In *Conference Proceedings, Research on Women's Issues in Transportation Volume 2: Technical Papers*. Transportation Research Board of the National Academies, Washington, D.C., 2004, pp. 31–40.
37. Emond, C. R., W. Tang, and S. L. Handy. Explaining Gender Difference in Bicycling Behavior. *Transportation Research Record: Journal of the Transportation Research Board*, 2009. 2125: 16–25.
38. Washington, S., M. Karlaftis, and F. Mannering. *Statistical and Econometric Methods for Transportation Data Analysis*. Chapman & Hall/CRC, Boca Raton, FL, 2003.
39. Jensen, S. U. Safety Effects of Blue Cycle Crossings: A Before-After Study. *Accident Analysis and Prevention*, Vol. 40, No. 2, 2008, pp. 742–750.
40. Weigand, L., N. McNeil, and J. Dill. *Cost Analysis of Bicycle Facilities: Cases from Cities in the Portland, OR Region*, Portland, OR, 2013. http://activelivingresearch.org/sites/default/files/Dill_Bicycle_Facility_Cost_June2013.pdf.
41. *Urban Bike Design Guide, April 2011 Edition*. National Association of City Transportation Officials, New York, NY, 2011. http://www.ocpcrpa.org/docs/projects/bikeped/NACTO_Urban_Bikeway_Design_Guide.pdf.

The Standing Committee on Transportation Planning Applications (ADB50) peer-reviewed this paper (18-01602).

Article

TRR
JOURNAL OF THE TRANSPORTATION RESEARCH BOARD

Transportation Research Record
2018, Vol. 2672(46) 55–63

DOI: 10.1177/0361198118782042
journals.sagepub.com/home/trr
SAGE

A German Passenger Car and Heavy Vehicle Stock Model: Towards an Autonomous Vehicle Fleet

Martin Hartmann[1] and Peter Vortisch[1]

Abstract

Automated vehicles are becoming a reality. Many pilot projects have already begun demonstrating the technological capabilities, as public authorities now allow the testing of automated vehicles in real traffic. To smooth the transition from a conventional to an automated fleet, effective fiscal and regulatory policies must be developed by governmental agencies. But at what rate will automated vehicles actually be adopted, and what automation technology will be available for use in new cars joining the national fleet? A national vehicle stock model can be used to answer these questions and to observe the aggregate impact of governmental policies on individual vehicle purchase decisions. In this paper, we present a passenger car and heavy vehicle stock cohort model that forecasts the diffusion of automation technology in Germany. The model uses national data on vehicle stock and vehicle utilization patterns on German freeways and predicts market shares of generic automation levels in predefined instances of a trend scenario. Results point toward market saturation of automated vehicles beyond 2050, with almost 90% of the passenger car fleet being classified as at least partially automatized by this date. The results also suggest that technology diffusion will be faster in the heavy vehicle fleet than in the passenger car fleet. This implies a positive correlation between emission-linked road user charges for heavy vehicles on the freeway network and the renewal rate of the heavy vehicle fleet. The forecast shares of automated vehicles can be used as an input for traffic flow simulations or as a basis for those infrastructure measures and traffic policies that are sensitive to the share of automated vehicles.

Rapid advances in vehicle automation technology are causing national governments to analyze future vehicle fleet scenarios in terms of developing rational automated vehicle (AV) policies. Car manufacturers are also interested in knowing the size and composition of future vehicle fleets in order to plan their production and sales for the medium and long term. Automated vehicle technology is expected to disrupt mobility behavior, however, which will shift trends in car ownership. New sharing concepts, around-the-clock available autonomous taxis, and a rapid increase in telecommuting are just a few examples of significant changes that will affect car ownership decision-making. At the same time, governmental fiscal and regulatory interventions and incentives can significantly speed up the introduction of new vehicle technologies and their subsequent market shares. A good example is the introduction of road user charging (RUC) for heavy vehicles in Germany, which significantly accelerated the renewal of the vehicle fleet in Germany and neighboring countries (*1*).

In this study, we present a data-based modeling approach for predicting the vehicle stock composition and the diffusion of new technologies into Germany's national vehicle fleet. It is based on historical data and current predictions about the vehicle stock and a forecast about the introduction of vehicle automation technology into the German market. The available data show that the vehicle fleet is very stable, with vehicle survival rates pointing towards mean life expectancies of more than two decades and an increasing share of vintage cars in the national fleet (*2*). Next, we derive long-term projections of the national vehicle fleet taking into account its longevity and identify the pace of vehicle technology diffusion. Arguably, there is a high level of uncertainty caused by numerous assumptions and a lack of knowledge about future governmental regulatory policies. However, the results can be viewed as "business-as-usual" estimates, in which current trends are projected forward. If desired, accelerated technology diffusion can be accounted for by incorporating governmental incentives into the model.

Our efforts are part of a research project that investigated the effects of vehicle automation on the freeway capacity in Germany (*3*). Hence, we limit our forecast of the automated vehicle share to the German freeway network, where we

[1]Institute for Transport Studies, Karlsruhe Institute of Technology (KIT), Karlsruhe, Germany

Corresponding Author:
Address correspondence to Martin Hartmann: Martin.Hartmann@kit.edu

expect automated vehicles to reach significant shares earlier than in more complex, urban road networks. The study covers the market penetration of both passenger cars (PC) and heavy vehicles (HV). We first offer a brief literature review of the existing national vehicle fleet models for Europe and the United States. We then provide a methodology overview and list the input data used for the vehicle stock modeling. The core of our paper details the components of the vehicle stock cohort model and shows resulting projections of vehicle automation technology diffusion in Germany. Lastly, we draw some conclusions about our results.

Literature Review

Estimating the composition and size of national vehicle fleets has been of interest ever since the car industry began to have a substantial impact on national economies. As environmental awareness grew, national vehicle fleet models were then also used to assess the impact of regulatory policies on the vehicle stock. Most recently, the introduction of alternative fuels and disruptive vehicle automation technology have once again made projections of the national vehicle fleet a priority of transportation researchers.

In Norway, where the government is spearheading European efforts to accelerate electric vehicle use by establishing substantial fiscal and regulatory incentives, a bottom-up stock-flow model was developed (*4*). Here, the authors provide example projections of low-carbon and business-as-usual scenarios, based on the forecasted renewal process of the Norwegian vehicle fleet.

In Germany, Kröger et al. developed a model that focuses on the diffusion of automated vehicle technology within the German passenger car fleet (*5*). The model forecasts technology diffusion for automation levels 4 and 5 (cf. SAE definition) through the year 2035. The target vehicle stock is divided into four size segments—small, compact, medium and large—and the automation technology is assigned to the vehicles using a top-down approach. The model yields the market shares of level 4 and 5 AVs in the vehicle fleet, along with the number of new vehicles for a trend scenario and an extreme scenario in 2035. According to their model, the total share of AVs is 17.4% in the trend scenario (level 4 = 7.3%; level 5 = 10.1%) and 42.4% in the extreme scenario (level 4 = 4.8%; level 5 = 37.6%). Plötz et al. (*6*) estimated the German market penetration of electric vehicles in their user-specific model based on real-world driving data. And, the VECTOR21 vehicle technology scenario model (*7*) estimated the market share of electric vehicles in Germany for multiple scenarios.

In the United States, a new model was proposed to make long-term forecasts of connected and automated vehicles (*8*). This model is based on a nation-wide survey of stated preferences about a possible AV purchase. Since the survey over-represented Texas and certain demographic classes, the survey sample was scaled using the 2013 American Community Survey's Public Use Microdata Sample. The authors make assumptions about the evolution of the vehicle technology price, consumer willingness to pay, and changes in governmental regulations. The model estimates approx. 25% level 4 penetration by 2045. Lavasani et al. developed an AV market penetration model based on the adaption patterns of previous disruptive technologies, such as cell phones and the internet (*9*). To do so, the authors used the generalized Bass diffusion model to investigate AV adoption rates on a household level. Their projections show 1.3 million AVs sold within five years from the assumed sales start in 2025, and predict market saturation around 2060, with approximately 87 million AVs sold.

In Sweden, a vehicle stock model was developed in 2006 for the Swedish Transport Administration to evaluate the impact of vehicle fleet policies (*10*). The model has three components: a car ownership model, a scrapping model, and a new-car purchase model. Recent comparison of the model forecasts with actual vehicle stock data showed significant discrepancies, which has led to on-going efforts to improve the model (*11*).

Methodology

Our model of automation technology diffusion within the German vehicle fleet has three components: a model of passenger car and heavy vehicle stock cohorts, a model of automation technology market introduction and penetration, and a model of vehicle utilization. Merging the first two components yields a model of vehicle turnover and vehicle technology diffusion with respect to trends in vehicle ownership and predicted technology adoption rates. The third component investigates the freeway utilization of both passenger car and heavy vehicle fleets and yields adjusted penetration rates of the vehicle stock into the German freeway network. Figure 1 shows the modeling workflow and the input data.

There is always a high level of uncertainty when new, disruptive technology enters a market. In our case, uncertainty stems from the fact that it is not only the automotive technology that will change, but also its utilization. Moreover, the vehicle stock and its utilization—especially in the heavy vehicle segment—is strongly linked to national economic performance and the legal framework regulating the freight market. In the past, there have been large fluctuations −33%; 15% in the sales of new heavy vehicles that closely mirror the fluctuations in German economic performance (*12*). We address this short-term market volatility by assuming conservative, long-term growth rates for both the Germany economy and that of neighboring countries.

We recognize that various automated vehicle classifications are available. Here, we use a generic model for partially- (PAV), highly- (HAV), and connected automated vehicles (CAV) for both the passenger car and heavy vehicle segments. We assume a segment-specific speed of diffusion for automation technology. For passenger cars, we expect automation to be a luxury feature, one that will not become

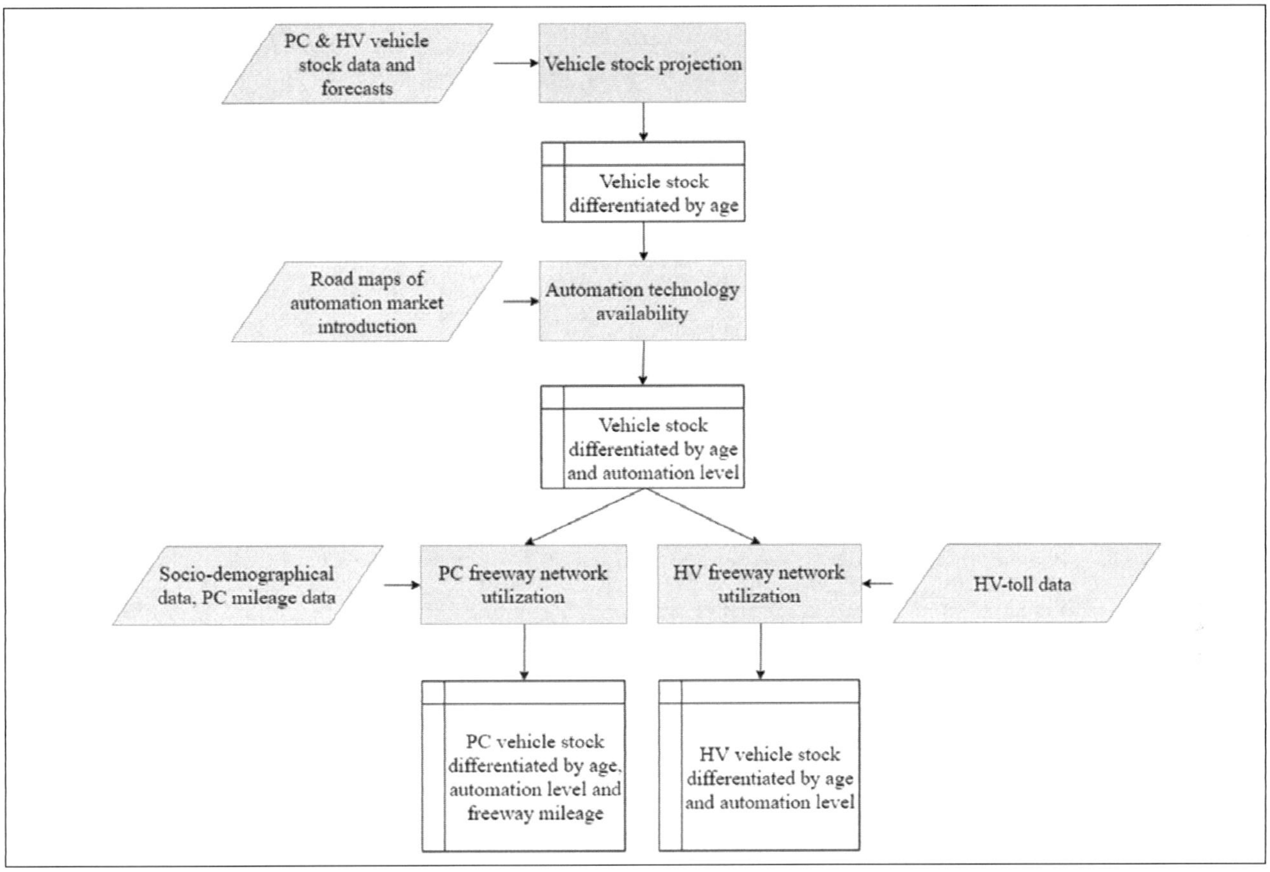

Figure 1. Flowchart of vehicle stock modeling in Germany.

mandatory in the near future. We therefore assume a diffusion of vehicle automation technologies following a top-down approach. For heavy vehicles, however, past experience shows that regulators tend to intervene strongly, often making the introduction of new vehicle technology mandatory (2004: ABS, 2014: ESP) within the fleet. This results in faster diffusion of automation within the heavy vehicle fleet, thanks to a faster pace of mandatory technology introduction into new, heavy-vehicle model years. Next, unlike road user charges (RUC) for passenger cars, heavy vehicles using European freeways are charged according to their emission class. This encourages freight operators to accelerate the renewal process of their fleets. Data from the German electronic RUC for heavy vehicles show that over 90% of the charged heavy vehicle fleet had been renewed within the previous 10 years to the strictest emission standard in effect at the time (2005–2015, Euro 5 and better).

As modeling input data, we use publicly-available administrative records and forecasts of the vehicle stock to create a vehicle stock database differentiated by age and year of first registration. Next, we review the announced time frames for technology introduction and its market availability, as announced by OEMs and automotive suppliers. We add historical data and consider past trends of the most important vehicle technologies (ABS, automatic transmission, navigation systems) to extract typical deployment cycles of technology diffusion. Through the modeled market availability of vehicle technology and shares of new automated vehicles, a self-sustaining platform is created in which the technology automatically diffuses through the cohorts as the vehicles age. Finally, we analyze the passenger car fleet in terms of freeway utilization to quantify the penetration rates of AVs in the vehicle stock on the freeway network.

A recognized limitation of the vehicle stock model is its inability to show secondary effects of vehicle automation on personal mobility. In particular, there are no rebound effects incorporated in the model. These include the impacts of shared mobility on the vehicle fleet (i.e., car sharing and e-hailing). Also, long-term effects on land use due to lower value of time of drivers travelling in automation-controlled convoys accepting longer travel times are not incorporated.

Model Components

Model 1: Vehicle Stock Cohort Model

Germany has a high level of motorization: In 2014, there were 547 passenger cars per 1,000 inhabitants, 3.03 million new passenger car registrations, 44.4 million passenger cars in total, and 2.89 million heavy vehicles (*13*). The records of vehicle stock and new registrations are collected by the Federal Motor

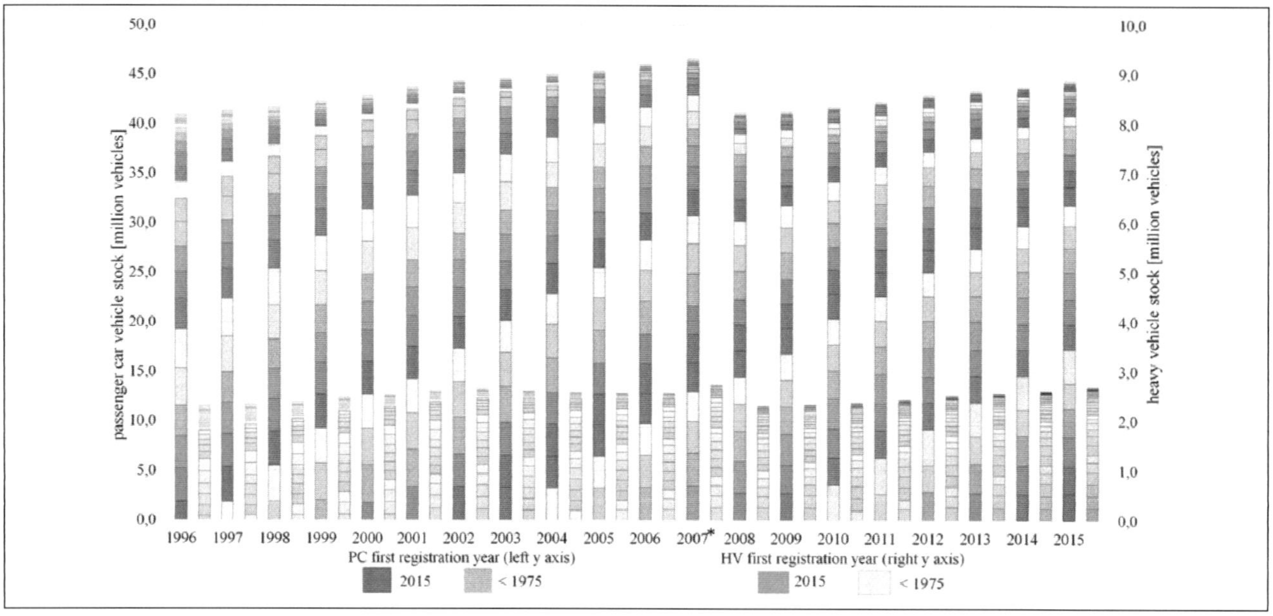

Figure 2. The German passenger car and heavy vehicle stock, clustered by 1996–2015 cohorts.

Transport Authority (KBA) and published annually (*12*, *2*). Figure 2 shows the development of the vehicle stock: The passenger car and heavy vehicle fleets have both grown by an average of 1% p.a. over the last two decades. The major drop depicted in 2008 was caused by modifications in the statistic methodology and does not influence the global trend itself. The saturation of the color scheme depicts the different vehicle cohorts and their share in the vehicle stock. The cohort's share and its gradual decrease within the vehicle stock reveals the vehicle model year's availability on the market and also the pace of the vehicle fleet renewal.

Based on the KBA data, we analyzed the annual aggregate characteristics of the vehicle fleet, as represented by the average vehicle age, the vehicle survival rate and life expectancy, and the number of new vehicle purchases. As this data shows, the average age of passenger cars increased by nearly one-third from 1996 (6.8 years) to 2015 (9.0 years) (*2*).

Due to the lack of official records on the vehicles exported from Germany, the statistic of mean survival rates of vehicles before scrapping has only limited explanatory power towards estimating the mean life expectancy of a model year. The average age of scrapped vehicles shows the life expectancy of the model year within the vehicle stock, without accounting for the impact of vehicle export or previous long-term decommissioning of the scrapped vehicles. Nevertheless, the mean survival rate before scrapping reaches 18 years (for comparison: 17.8 years in Norway (*4*)) in the passenger car segment and 25 years in the heavy vehicle segment (*14*). Table 1 summarizes the analysis of the vehicle stock records between 1996 and 2015.

Germany is home to a growing number of vintage cars, which are defined as vehicles older than 30 years that are kept in their original manufactured conditions. The stock of vintage cars reached about half a million in 2015—some 1% of the entire fleet (*2*). However, since they are, per se, not candidates for automation, we do not consider them further in this study.

Finally, we compiled annual projections of the passenger car and heavy vehicle stock through 2050. Here, we extrapolated the historical vehicle stock data given in Figure 2 and constrained the annual aggregated indicators of the forecasted vehicle stock to reflect those published in two independent studies (*15*, *16*). Both studies estimated saturation of the passenger car market around 2024 (so-called "peak car"). The resulting projections and the market saturation are illustrated in Figure 3. The 40 vehicle cohorts from the KBA data set (1975–2014) are displayed in color, and the 36 forecasted vehicle cohorts (2015–2050) are displayed in grayscale. The projection shows moderate growth in the heavy vehicle fleet through 2050, and a reduction in the passenger car stock after peak car is reached in 2024. By 2050, the passenger car fleet is expected to have shrunk in size (for the same reasons leading to peak car) to that observed in the 1990s.

Model 2: Availability of Automation Technology in New Model Year Vehicles

The pace of diffusion of technological advances into the vehicle fleet depends on how quickly the technology is introduced into new vehicles and on the market share of the equipped vehicles. Today's new vehicles can be equipped with assistance systems involving active lateral and longitudinal control, such as Lane Keeping Assist, Traffic Jam Assist, and Adaptive Cruise Control (ACC). In the near future, level 3 automation, such as Highway Chauffeur, will

Table 1. Aggregate Indicators of the Vehicle Stock in Germany (1996–2015).

Indicator	Unit	Passenger cars	Heavy vehicles
Average vehicle age (2015)	[yrs]	9.0	7.7
Average age increase p.a.	[yrs]	0.14	0.03
Life expectancy before scrapping	[yrs]	18	25
New vehicle sales (2015)	[1.000 veh]	2,762	264
New vehicle sales volatility (1996–2015)	[%]	[−23%, 23%]	[−33%, 17%]

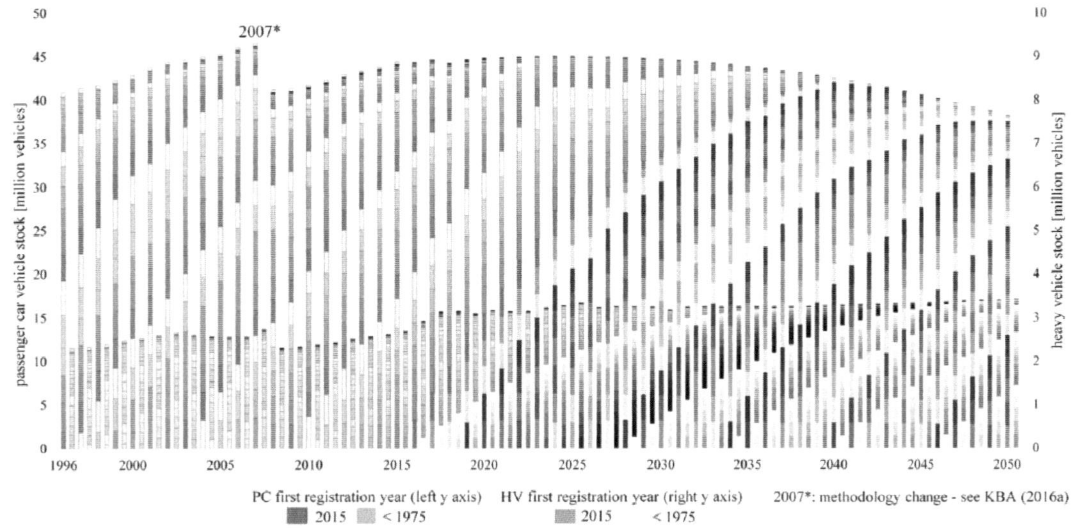

Figure 3. Projection of the vehicle stock in Germany, clustered in PC and HV cohorts.

allow un-supervised driving under some traffic conditions, so that drivers will not have to remain fully engaged in the driving task. When level 4 and level 5 technology is eventually introduced, vehicles will then operate independently within their design domains under all traffic conditions. This will lead to changes in typical mobility patterns and traditional vehicle ownership models.

To forecast future shares of automated vehicles within the vehicle fleet, we reviewed literature (*17*, *18*) and OEM roadmaps (*19*) and incorporated the published results into our own estimates of AV technology introduction in new vehicles. We selected the following assistance systems as being representative of automation technology and assigned them to the generic automation levels in the following manner:

- partially automated vehicles (PAV): ACC, Traffic Jam Assist, Lane Change Assist
- highly automated vehicles (HAV): PAV-functions + Highway Chauffeur (reactive)
- connected highly automated vehicles (CAV): PAV-Functions + Cooperative ACC + Highway Chauffeur (proactive)

Next, we adjusted the pace of technology diffusion to that observed in the past, assuming the ABS and ESP introductions to be typical for the passenger car fleet. From its introduction in 1978, ABS took 20 years to reach 40% market penetration in Western Europe; ESP required 10 years (*20*). This slow pace of market penetration was observed before the technology was made compulsory in 2004 and 2014, respectively, which illustrates the high inertia of the vehicle fleet renewal process. The sluggish tempo of the vehicle fleet turnover is confirmed by the long-term trends in European transport (*21*) and by the evidence derived from environmental policy actions (*22*). Figure 4 shows projections of technology availability in new model years, constrained to an average pace of past vehicle technology diffusion (based on ABS and ESP).

Model 3: Analysis of Vehicle Utilization on German Freeways

Due to the higher complexity of urban traffic conditions, we expect automated driving to first be deployed within the freeway network. Ergo, today's policy-makers are especially interested in identifying the challenges of freeway operation, particularly in mixed traffic conditions. Therefore, in the

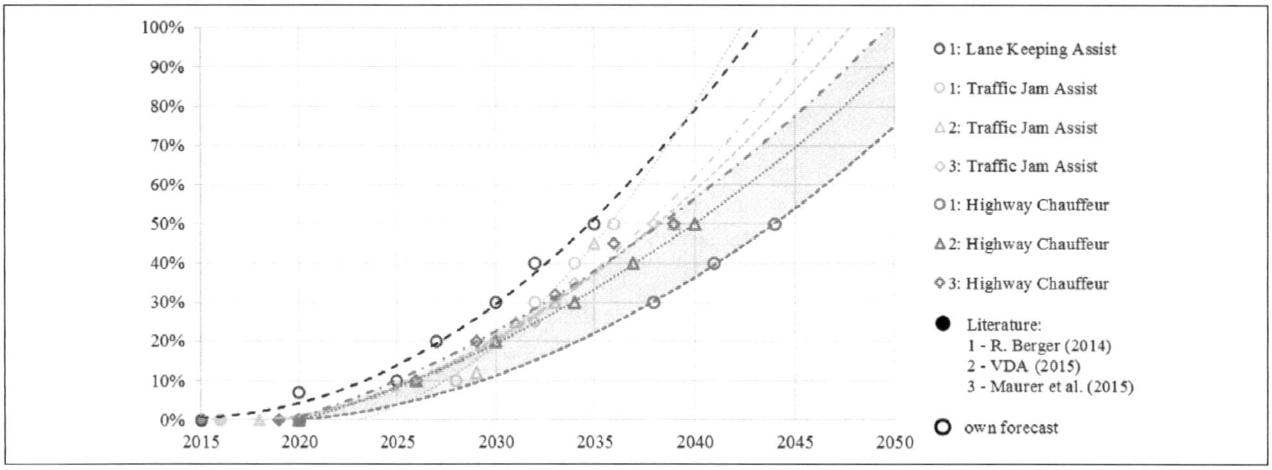

Figure 4. Projected market shares of vehicle technology in new model years.

third component of our model, we analyze the utilization patterns of passenger cars and heavy vehicles on the freeway network. The question to be answered here is, how much greater will the share of automated vehicles be on freeways than on the rest of the road network?

Analysis of Freeway Utilization by Passenger Cars. To quantify the additional share of AVs found on the freeway network, we first took a representative set of German households and—using a newly designed heuristic algorithm—assigned AVs to the household vehicle fleet to simulate a fictive AV purchase. As a basis for these assignments, we used data from the household travel survey "German Mobility Panel" (MOP) (*23*). We then analyzed the driving patterns of the households with AV vehicles to differentiate between freeway and non-freeway driving. This allowed us to derive the extra miles driven by AV vehicles on the Germany freeway network.

The dataset consists of 4551 representative German households in which weekly mobility patterns were reported for three years. For details on the survey's methodology, we refer to the final report (*23*). In our paper, data are extracted about such household determinants as the number of vehicles in the household, the annual miles driven on and off freeways, and household income. The annual average mileage of all vehicles in the sample adds up to 8000 miles; the mileage driven on freeways amounts to 3000 miles. Our heuristic algorithm clusters the households based on the above-described determinants and assigns them to multiple instances of a trend scenario between 2015 and 2050 and beyond (2050+). Based on the assumption that high-income households with high annual vehicle mileage will be quicker to adopt automated vehicles, thresholds were set incrementally to represent a household's willingness to purchase an AV. Finally, the model was constrained by theoretical adoption shares derived from a distribution based on the (technology) adoption theory (*24*). Table 2 shows the assignment results, along with their deviations from the target values based on the adoption theory.

The resulting additional shares vary for the different automation levels and the instances of the trend scenario. In a trend scenario—with no future limited freeway-access policy—there is an estimated average of 10% more automated vehicles on German freeways than on the rest of the road network. We therefore add this estimated additional share of AVs to the share of AVs computed in component models 1 and 2 to account for increased freeway utilization. Figure 5 shows the resulting projection of automation diffusion. The interpretation of the results shows that the level of automation depends on the number of new vehicles sold and on the technological equipment found in these autos. Although the highest automation level in this study, CAV, should dominate the vehicle stock in the distant future, there will be enough partially automated economy vehicles purchased in the next years so that their share in the vehicle stock remains significant until 2050.

Analysis of Freeway Utilization by Heavy Vehicles. The official RUC statistics collected on the entire freeway network represent a rich aggregate database about heavy vehicle use in Germany (*1*). A high proportion (40%) of the registered heavy vehicles operating on German freeways are non-German. In 2015, there were about 1.5 million heavy vehicles that were charged at least once, representing 18.5 billion charged miles. Only about 30 % of the German heavy vehicle stock uses the freeway network at least once a year. This supports the assumption that a fairly minor share of the heavy vehicle stock accounts for nearly the entire distance traveled by heavy vehicles on German freeways.

As previously mentioned, the large percentage of charged heavy vehicles meeting the strictest emission standards at

Table 2. Heuristic-Based Assignment of Automated Vehicles to Representative German Households.

Threshold attribute	Automation level	2015	2020	2025	2035	2045	2050+
Annual vehicle mileage [1.000 mi]	*CAV*	>25,0	>12,5	>10,5	>8,8	>6,3	>5,5
	PAV/HAV	>18,7	>9,5	>6,3	>5,5	>2,5	>1,8
	AND						
Number of passenger cars per household	*CAV*	>2	>2	>2	>0	>0	>0
	PAV/HAV	>1	>1	>1	>0	>0	>0
	AND						
Household income [1.000 €]	*CAV*	>3,5	>3,5	>3,5	>3,0	>3,0	>3,0
	PAV/HAV	3,0–3,5	3,0–3,5	2,0–3,5	1,5–3,5	1,0–3,5	0,5–3,5
Assignment results	*CAV*	0%	1%	6%	12%	21%	25%
	PAV/HAV	0%	2%	10%	32%	55%	60%
	Conventional	100%	97%	84%	56%	24%	15%
Adoption theory	*Σ assigned fleet*	0%	3%	16%	44%	76%	85%
	Target value	0%	3%	16%	50%	84%	90–100%
	Difference	0%	0%	0%	6%	8%	5–15%

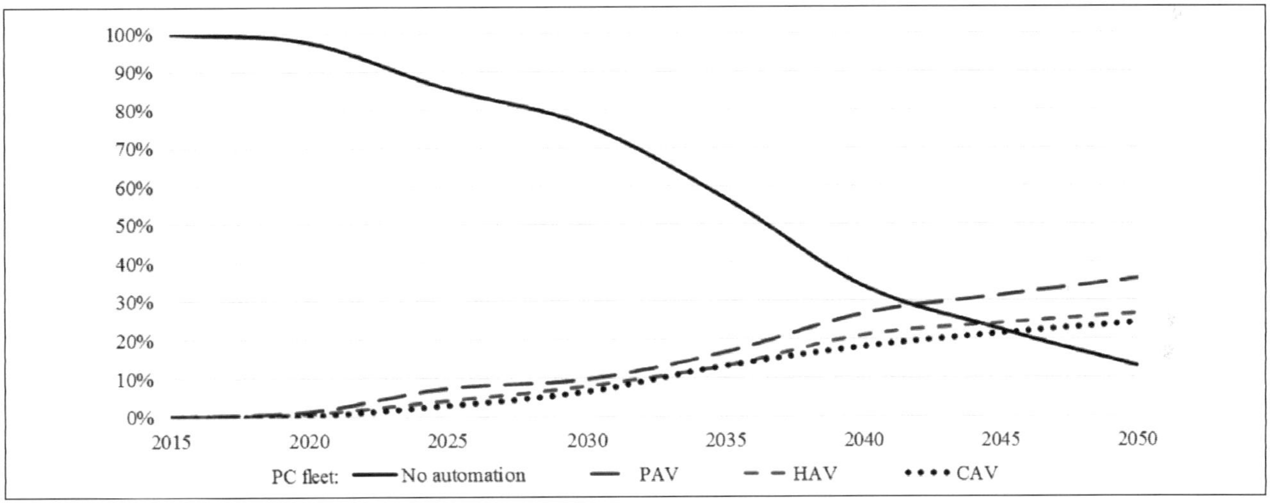

Figure 5. Projection of automation diffusion within the German PC fleet until 2050.

their time of purchase indicates a fast pace of renewal for the charged heavy vehicle fleet. The RUC statistic further reveals that there is no major discrepancy between the emission class of the German heavy vehicle fleet and the fleets of other European countries. This suggests a comparable renewal pace of foreign fleets. One might thus conclude that automation diffusion will be comparable in German and non-German heavy vehicles on the charged roads in Germany. Generally speaking, the heavy vehicle fleet on German freeways is represented by a modern, rapidly renewing, and comparatively homogenous sample of the entire heavy vehicle stock in Germany. Thus, there is potential for rapid introduction of vehicle automation within the heavy vehicle segment subjected to RUC.

In this study, we derive the market shares of automated heavy vehicles based on the shares of the German and foreign heavy vehicle stock and the availability of vehicle technology within the heavy vehicle fleet. This is analogous to our approach for the passenger car segment. Since the RUC is collected solely on the highway network, we do not need to estimate additional shares of heavy vehicle AVs on the freeways. As shown in Figure 6, the accelerated technology diffusion resulting from mandatory technology introduction in new heavy vehicles and the impact of RUC leads to a 60% share of at least partially automated heavy vehicles by 2030 (vs. 25% for passenger cars). According to our projections, the share of non-automated heavy vehicles on freeways will be insignificant by around 2045.

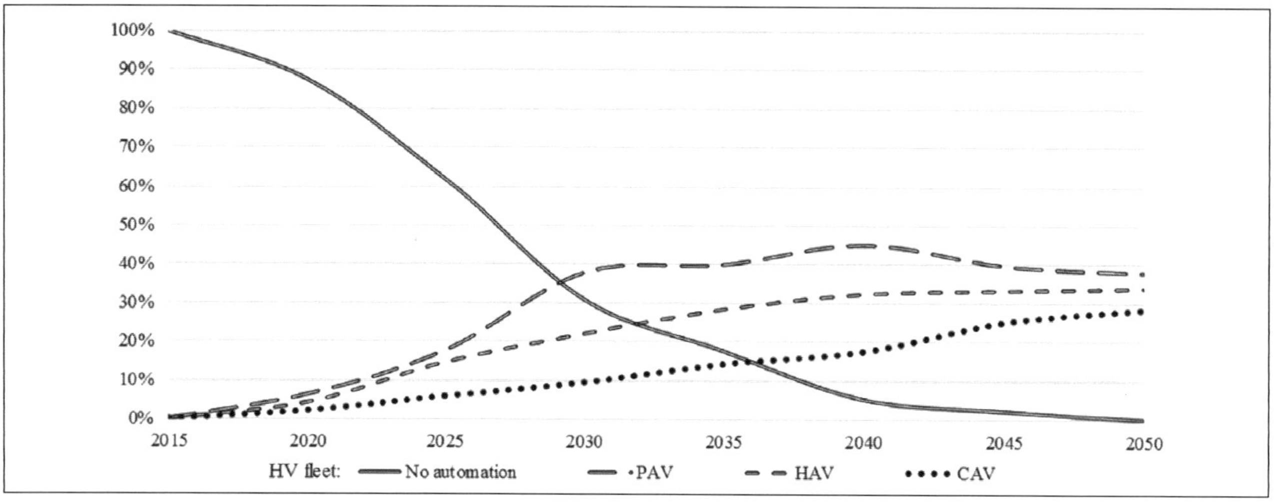

Figure 6. Projection of automation diffusion within the German HV fleet until 2050.

Conclusion

At the dawn of major disruptions in the automotive industry, which are being driven by digitalization and the sharing economy, vehicle automation is generating increasing interest among policy makers and the major players in the automotive market. The vehicle stock model presented here represents one of the first attempts to forecast the number of automated vehicles that will be operating on German freeways in the medium term and long term. This information is especially important for examining potential infrastructure and transportation policy measures before their implementation, since they are affected significantly by the share of automated vehicles in the overall vehicle fleet. Alternatively, the model causality could be reversed, thus allowing decision makers to investigate what policy measures would be necessary to reach desired AV shares within a certain time horizon.

Examining the vehicle stock yields insights into the dynamics of fleet renewal and reveals the high level of inertia in the German vehicle fleet. This correlates with vehicle stock data showing that it takes major regulatory interventions, such as emission-class-sensitive RUC, national directives for mandatory technology introduction, or temporary scrapping incentives to significantly boost new vehicle sales and thus accelerate the fleet penetration rates of vehicle technology.

The resulting projections give conservative estimates of technology diffusion for a trend scenario that extends to 2050. The projected AV shares (e.g., 11% of CAV in 2035) are comparable to the results of a similar vehicle technology diffusion study in Germany (*5*), where the "Trend Scenario 2035" indicated a 10% share of level 5 vehicles. By 2050, however, our study predicts a 90% share of at least partially automated passenger cars and a 100% share of heavy vehicles. Thus, one can expect a slow-paced, but steady penetration of automation technology within the vehicle fleet.

It is generally argued that unforeseeable, technology-triggered regulations will significantly affect technology diffusion. Once an average automated vehicle becomes statistically safer than an average human driver, for example, policy makers will undoubtedly consider the ethical implications of this situation in legislation and might even ban conventional vehicles from the operational design domain of the automated vehicles. This, in turn, would significantly accelerate AV penetration. Since this kind of disruptive policy measure is difficult to convert into a reliable and robust mathematical statement, the trend scenario modeled here only considers conservative, non-disruptive regulatory policies.

Acknowledgments

This paper is based on the research sponsored by the Research Association of Automotive Technology (FAT)—working group Optimization of Road Traffic Systems—as a member of the German Association of the Automotive Industry (VDA). The authors confirm contribution to the paper as follows: study conception and design: M. Hartmann, P. Vortisch; data collection: M. Hartmann; analysis and interpretation of results: M. Hartmann, P. Vortisch; draft manuscript preparation: M. Hartmann. All authors reviewed the results and approved the final version of the manuscript.

References

1. BAG. *Heavy Vehicles Road Toll Statistic*. Annual Evaluation 2015, Cologne, 2016.
2. KBA. (ed.). *Registrations of Passenger Cars and Heavy Good Vehicles in Germany Distinguished by Age (FZ15)*. Kraftfahrt-Bundesamt - Federal Motor Transport Authority, Flensburg, 2016.
3. Krause, S., N. Motamedidehkordi, S. Hoffmann, F. Busch, M. Hartmann, and P. Vortisch. *Auswirkungen des teil- und hochautomatisierten Fahrens auf die Kapazität der Fernstraßeninfrastruktur*. Forschungsvereinigung Automobiltechnik e.V, Berlin, 2017, http://dx.doi.org/10.13140/RG.2.2.20460.56968.

4. Fridstrøm, L., V. Østli, and K. W. Johansen. A Stock-Flow Cohort Model of the National Car Fleet. *European Transport Research Review*, Vol. 8, No. 3, 2016, p. 1. http://dx.doi.org/10.1007/s12544-016-0210-z.
5. Kröger, L., T. Kuhnimhof, and S. Trommer. Modelling the Impact of Automated Driving. Private AV Scenarios for Germany and the US. *Proc., European Transport Conference*, Barcelona, 2016. http://elib.dlr.de/110333/.
6. Plötz, P., T. Gnann, and M. Wietschel. Modelling Market Diffusion of Electric Vehicles with Real World Driving Data — Part I: Model Structure and Validation. *Ecological Economics*, Vol. 107, 2014, pp. 411–421. http://dx.doi.org/10.1016/j.ecolecon.2014.09.021.
7. Propfe, B., D. Kreyenberg, J. Wind, and S. Schmid. Market Penetration Analysis of Electric Vehicles in the German Passenger Car Market towards 2030. *International Journal of Hydrogen Energy*, Vol. 38, No. 13, 2013, pp. 5201–5208. http://dx.doi.org/10.1016/j.ijhydene.2013.02.049.
8. Bansal, P., and K. M. Kockelman. Forecasting Americans' Long-Term Adoption of Connected and Autonomous Vehicle Technologies. *Transportation Research Part A: Policy and Practice*, Vol. 95, 2017, pp. 49–63. http://dx.doi.org/10.1016/j.tra.2016.10.013.
9. Lavasani, M., X. Jin, and Y. Du. Market Penetration Model for Autonomous Vehicles on the Basis of Earlier Technology Adoption Experience. *Transportation Research Record: Journal of the Transportation Research Board*, 2016. 2597: 67–74.
10. Transek. *Bilparksmodell (Car Fleet Model)*. Technical Report. Transek AB, Sweden, 2006.
11. Beser Hugosson, M., S. Algers, S. Habibi, and P. Sundbergh. Evaluation of the Swedish Car Fleet Model using Recent Applications. *Transport Policy*, Vol. 49, 2016, pp. 30–40. http://dx.doi.org/10.1016/j.tranpol.2016.03.010.
12. KBA. (ed.). *Newly Registered Vehicles in Germany*. Kraftfahrt-Bundesamt - Federal Motor Transport Authority, Flensburg, 2016.
13. BMVI. (ed.). *Verkehr in Zahlen 2016/2017*. DVV Media Group GmbH, Hamburg, 2016.
14. WirtschaftsWoche. (n.d.). *Typical Vehicle Survival Rates in Germany by Vehicle Brand. In Statista - Das Statistik-Portal*, 2014. https://de.statista.com/statistik/daten/studie/316498/umfrage/lebensdauer-von-autos-deutschland.
15. Intraplan Consult GmbH. *Verkehrsverflechtungsprognose 2030: Zusammenfassung der Erbenisse*. Bundesministerium für Verkehr und digitale Infrastruktur, Berlin, 2014.
16. Shell Deutschland, P. A. G. *Shell PKW-Szenarien bis 2040*. Shell Deutschland Oil GmbH, Hamburg, 2014.
17. Bernhart, W. *Autonomous Driving: Think Act. Roland Berger Strategy Consultants GmbH, München*, 2014. https://www.rolandberger.com/de/Publications/pub_autonomous_driving.html.
18. Maurer, M., J. C. Gerdes, B. Lenz, and H. Winner. *Autonomes Fahren*. Springer-Vieweg, Wiesbaden, 2015.
19. VDA. *Schritt für Schritt in die Zukunft* [Online]. https://www.vda.de/de/themen/innovation-und-technik/automatisiertes-fahren/schritt-fuer-schritt-in-die-zukunft.html, 2015.
20. Gottschalk, B., R. Kalmbach, and J. Dannenberg. *Markenmanagement in der Automobilindustrie: Die Erfolgsstrategien internationaler Top-Manager*. Springer-Verlag, Heidelberg, 2015.
21. Moraglio, M., and H.-L. Dienel. Shifts, Turning Points and Inertia Exploring Long-Term Industry Trends in European Transport. *European Journal of Futures Research*, Vol. 3, No. 1, 2015, p. 11. http://dx.doi.org/10.1007/s40309-015-0070-0.
22. Ghersi, F., and S. McDonnell. The Impacts of Long-Term CO2 Objectives on Short-Term Transportation Trends in the European Union. *Energy for Sustainable Development*, Vol. 11, No. 3, 2007, pp. 33–43. http://dx.doi.org/10.1016/S0973-0826(08)60575-X.
23. Weiß, C., B. Chlond, T. Hilgert, and P. Vortisch. *Deutsches Mobilitätspanel (MOP) - Wissenschaftliche Begleitung und Auswertungen. Bericht 2014/2015: Alltagsmobilität und Fahrleistung*. KIT Institut für Verkehrswesen, Karlsruhe, 2016.
24. Rogers, M. *Diffusion of Innovations*, 5th ed. Free Press, New York, 2003.

The Standing Committee on Transportation Planning Applications (ADB50) peer-reviewed this paper (18-04389).

Article

Transportation Research Record
2018, Vol. 2672(46) 64–72

DOI: 10.1177/0361198118798456
journals.sagepub.com/home/trr

The Pros and Cons of Using the Change in Destination Choice Logsums as a Practical Measure of User Benefits

Kristen Villanueva[1], Lisa Zorn[2], David Ory[3], and David Vautin[2]

Abstract

The Metropolitan Transportation Commission carries out an extensive project performance assessment in support of regional planning efforts. For the 2017 regional transportation plan, various types of potential transportation investments were evaluated against qualitative and quantitative criteria. The quantitative metric was a benefit–cost ratio, in which a project's expected annual benefits were compared with the project's expected annualized cost. To quantify user benefits (one component of the total benefit estimate), agency staff used the difference in Project/No Project destination choice logsums generated by the region's activity-based travel model. The destination choice models include, as covariates, mode choice logsums, which means the change in destination choice logsums estimates the change in consumer surplus across all destinations and all travel modes. This approach has long been advocated for in academia and was applied by the Federal Transit Administration for evaluating transit investments in the 2000s. This is the first time that it has been fully implemented in a large-scale, multi-modal application that informs investment decisions. This paper discusses the pros and cons of using this approach. The pros include consistency between simulated behavioral changes in the travel model and quantification of benefits, which cannot be achieved using typical measures of benefits such as travel time savings, as well as the ability to map changes in access across the region. The cons include issues related to the form and creation of alternatives in the mode choice model and the difficulty in explaining changes in logsums to non-technical audiences.

For the long-range regional transportation plan (Plan) for the San Francisco Bay Area, major investments are prioritized based in part on evaluations of cost-effectiveness and qualitative support of Plan targets. Projects that are found to be cost-effective and supportive of targets are candidates for discretionary funding in the Plan. Projects that are not cost-effective or do not support Plan targets undergo a separate process to appeal their low-performing status to the governing board. Consequently, many low-performing projects do not move forward into the fiscally constrained Plan (*1*). Given the emphasis on performance in prioritizing investments, accurate and robust evaluation methods are crucial.

One of the key inputs to cost-effectiveness is the change in benefits that travelers experience after a project is constructed. These "user benefits" are commonly estimated with changes in travel time and out-of-pocket costs to the traveling public. Increasingly, regions are relying on their activity-based travel models to estimate benefits of transportation investments. This paper presents one such application of estimating user benefits as the change in destination choice logsums, a measure of consumer surplus and accessibility, through the region's activity-based travel model. The paper presents the context of the application and decision-making framework, explains the theoretical relevance of logsums to project evaluation, and presents the pros and cons of the practical application for the San Francisco Bay Area region.

Background

The performance assessment described in this paper was conducted to help prioritize investments for the San Francisco Bay Area's Plan, which is developed by the Metropolitan Transportation Commission (MTC) every

[1]Alameda County Transportation Commission, Oakland, CA
[2]Metropolitan Transportation Commission/Association of Bay Area Governments, San Francisco, CA
[3]WSP, Montpelier, VT

Corresponding Author:
Address correspondence to Kristen Villanueva:
kvillanueva@alamedactc.org

4 years. The planning process for the Plan began in 2015, and the Plan itself was finalized and adopted in July 2017 (*2*).

At the beginning of the planning cycle, regional agencies forecast an increase of nearly one million new households through 2040 and approximately US$400 billion in transportation funding needs for operations, maintenance, modernization, and expansion. Available funding was estimated at only US$303 billion, leaving agency staff and the region to prioritize major investments.

Starting with the 2009 Plan (*3*) and significantly expanded in the 2013 Plan (*4*), regional decision-makers and staff prioritized investments by relying in part on a quantitative benefit–cost assessment for major investments (e.g., investments over US$100 million). The benefit–cost assessment consolidates a suite of outputs from the region's travel model into one metric through monetization, which is both easy to communicate and by virtue of being administered by agency staff, can be considered an independent project appraisal. What also differentiates this assessment from others is that a single benefit–cost methodology is applied consistently across investments, regardless of project type, to enable a uniform comparison of projects. The 2017 Plan, *Plan Bay Area 2040*, continued the project performance framework of the previous plans with a focused update of the user benefit methodology.

Performance and Modeling Context

The authors evaluated benefit–cost ratios for nearly 70 different investments in preparation of *Plan Bay Area 2040*. These investments ranged from rail extensions to construction of new high occupancy toll lanes. Different levels of maintenance investments were also evaluated for highways, local streets and roads, and public transit. Detailed documentation on the methodology for evaluating these "state of good repair" investments is available (*5*, *6*) with additional description on the overall performance assessment in *Plan Bay Area 2040 Final Supplemental Report: Performance Assessment Report* (*7*).

The benefit–cost calculation monetizes project effects on travel time, travel cost, emissions, collisions, health, and noise, detailed categories for which are summarized in Table 1 (*7*). The monetary value of travel time, for example, is related to the average regional wage rate, and the monetary value of particulate matter emissions reflects the costs associated with its known health impacts (*1*, *7*). To complete the assessment, annual life-cycle costs are estimated, reflecting any revenues expected from the project, and used as the denominator in the benefit–cost ratio.

The primary tool for forecasting benefits is the region's activity-based travel model, known as Travel Model One. For all projects and maintenance investments, each project was coded as its own "Project" scenario and compared with a "No Project" baseline, which reflected the existing transportation network plus projects under construction in 2015. Maintenance projects were treated differently: the No Project baseline used to compute these benefits was an extreme scenario that assumed zero maintenance funding. Both the Project and No Project scenarios reflected the land use assumptions in the 2013 Plan (*4*) and were run only for the horizon year of 2040. Each model run consisted of three iterations using sampling rates of 5%, 10%, and 50%, respectively. Every component of the travel model was allowed to vary in each iteration, including long-term (e.g., automobile ownership, workplace location) and short-term (e.g., mode choice) travel decisions.

Table 1. Benefit Categories Captured in the Benefit–Cost Ratio

User benefits: Travel time and cost
- Passenger vehicle travel time and operating cost
- Commercial vehicle travel time and operating cost
- Non-recurring freeway delay
- Transit in-vehicle travel time
- Transit out-of-vehicle travel time
- Vehicle ownership
- Parking cost
- Walk/bike travel time
- Transfer correction: fares, tolls

Society benefits: Air quality
- $PM_{2.5}$ (tailpipe gasoline, tailpipe diesel, road dust, and brake and tire wear)
- NO_X and SO_2
- VOC (acetaldehyde, benzene, 1,3-butadiene, formaldehyde, and others)
- CO_2

Society benefits: Safety and health
- Fatalities due to collisions
- Injuries due to collisions
- Property-damage-only collisions
- Reduction in disease from active lifestyles (morbidity)
- Estimated deaths averted from active lifestyles (mortality)
- Noise

Note: $PM_{2.5}$ = particle matter in the air smaller than or equal to 2.5 microns; NO_x = nitrogen-oxygen compounds; SO_2 = sulphur dioxide; VOC = volatile organic compounds; CO_2 = carbon dioxide.

To estimate the benefits listed in Table 1, Travel Model One output regional travel and operational metrics such as vehicle miles traveled (VMT) and time and costs spent traveling by different modes. The authors developed a set of scripts to compare the Project- and No-Project scenario output metrics, convert them into terms that could be monetized, apply the latest industry standard for valuations (e.g., value of time, statistical value of a life, etc.), and annualize the results. The set of scripts are open source and available on GitHub as part of Travel Model One (*8*).

Performance results were shared with project sponsors through a performance dashboard, available here: http://bayareametro.github.io/performance/dashboard/. The highest performing projects, those with medium-high benefit–cost ratios and medium-high qualitative target scores, were prioritized for funding endorsements such as for the Federal Transit Administration's (FTA) New/Small Starts and the state's Cap and Trade programs as well as for regional discretionary funding such as future increases in bridge tolls.

Motivation for a New User Benefit Methodology

MTC's performance work for *Plan Bay Area* (the 2013 Plan) used the region's activity-based model to estimate user benefits of a project by comparing the Project scenario's total travel time and total cost by mode with that of the No Project scenario, and monetizing the difference (*9*). We will refer to this methodology as the "realized travel time and cost" method since it used travel metrics for trips "taken" during the model simulation. As this was the first time the region's activity-based model was used to evaluate nearly 80 different individual investments, the authors noticed a clear inconsistency between the model's utility maximizing framework and the economic valuation of user benefits, particularly for transit investments. Projects that stimulated mode-switching were generating low or negative economic benefits when differencing total travel time and cost. This is a known issue: when travelers maximize their utility in the travel model, there are often circumstances in which a traveler may choose a slower option to realize other benefits. Furthermore, the region's activity-based model is stochastic, meaning there is a probabilistic response to the introduction of a change in transportation supply.

To illustrate this point, consider the following hypothetical example: for a particular origin and destination, 100 people can travel by either car or bus to get to work. Driving takes 10 min and costs US$1 whereas the bus takes 20 min and costs US$1. In this base scenario, 70 people drive and 30 people take the bus. Now suppose the bus is upgraded to a bus rapid transit (BRT) facility, meaning the bus now travels faster and more frequently but still costs the same. With the project, the bus trip now takes 15 min (5-min faster), which encourages 20 people to switch from driving to taking the bus by virtue of seeing an improvement in access and by virtue of the stochastic nature of the activity-based model. Compared with the baseline case, travel times for those mode-switchers is 5-min slower. Monetizing the difference in travel times would suggest there is an economic cost (or disbenefit) associated with choosing the BRT option. This outcome is counter to the premise that people change modes to maximize their utility and thus should be better off. To accurately reflect the benefit of a transportation project, especially one that induces mode-switching, the performance assessment needed to incorporate a user benefit methodology that took this into account.

To minimize invalid disbenefits from mode-switching, the methodology for *Plan Bay Area* (the 2013 Plan) included an adjustment to the estimate of out-of-vehicle transit travel time (*9*, p. 36), acknowledging that out-of-vehicle travel time (OVTT) was the main difference between the estimates of travel time for automobile travel and transit. The adjustment involved strategically reducing increases in OVTT between scenarios based on trips likely to have switched modes. In the performance assessment for *Plan Bay Area 2040* (the 2017 Plan), the authors sought a more elegant solution to this limitation of focusing on realized travel time and travel cost.

User Benefits with Destination Choice Logsums

When updating the user benefit methodology for *Plan Bay Area 2040*, the authors turned to literature on applied welfare economics for a stronger connection between the discrete choice equations of the activity-based travel model and user benefits for project appraisal. Fundamentally, user benefits estimate the change in welfare of the traveling population, which is also referred to as the change in consumer surplus. The seminal works of Williams (*10*) and Small and Rosen (*11*) are frequently referenced in papers describing the theoretical basis for estimating consumer surplus in the discrete choice context. These papers establish the connection between the estimation of demand curves for travel with consumer surplus after a few transformations. This is particularly straightforward if the coefficient on income is constant in the choice equations of the travel model, meaning the marginal utility of income is constant (*12*).

De Jong et al. present an in-depth review of using logsums in theoretical project evaluations as well as describe seven practical applications since 2000 (*12*). The authors also present a case study examining the differences in benefits of a rail improvement for the method with realized travel time and cost changes and a tour-based logsum method (using two methods for monetization of the logsum metric). The authors conclude that some advantages of the logsum method include ease of calculation (the model already estimates logsums for its estimate of demand) and ability to estimate user benefits when a new alternative is introduced. Kohli and Daly offer that an additional benefit of the logsum method is its ability to estimate the distributional benefits of transportation investments (*13*).

One practical application relevant to the application in this paper is the use of a logsum-based estimate of user benefits for the FTA's annual New Starts Application

(*14*). Referenced in De Jong et al. (*12*), Freedman, Castiglione, and Charlton (*15*) describe an evaluation of a subway extension in downtown San Francisco, California, using San Francisco County's activity-based model, SF CHAMP. At the time, the FTA methodology required that "total trips by zone pair and market segment remain fixed between the baseline and build alternatives." The authors note that this constraint is difficult to apply in the activity-based context. Introducing a significant change in transport supply (and associated accessibilities) influences household vehicle availability and tour generation, which then affects the frequency of intermediary trip generation and mode, leading to an inconsistency between the FTA methodology and an unconstrained assessment of user benefits from an activity-based model.

More recently, San Diego County has also begun to incorporate estimates of benefits of individual projects using their activity-based model in their long-range planning (*16*). Subsequently, the Federal Highway Administration funded research testing different types of logsum-based measures for estimates of user benefits. Resource Systems Group (RSG) describes multiple tests of two activity-based models: one for Tampa and one for San Diego (*17*). The report concludes that "accessibility logsums" promise to be a straightforward, practical application of estimating consumer surplus for agencies with both activity- and trip-based models. The report defines accessibility logsums as "generated by joint destination and mode choice logsums," which are pre-calculated for all residence zones and market segments. Some primary advantages include capturing benefits of mode-switching and destination-changing, handling cases when modes change in level of availability between the base and build scenarios, and providing a stable estimate throughout random variation between model runs/iterations.

The application for *Plan Bay Area 2040* described in this paper builds on numerous theoretical examples and a modest number of practical examples of using logsums to evaluate user benefits. This application goes further than most examples and inserts destination choice logsums as the primary component of the benefit–cost assessment for nearly 70 transportation investments, the results of which were extensively vetted in a public process throughout development of a long-range Plan. Note that this paper uses the term "destination choice logsum" to refer to the same concept as the RSG report's "accessibility logsum" (*17*).

Implementation of Destination Choice Logsums in Travel Model One

The estimate of consumer surplus in Travel Model One involved the difference between Project/No Project destination choice logsums as the change in accessibility for travelers throughout the region. Figure 1 shows the classic diagram illustrating consumer surplus but within the travel model context. Price is the independent variable and can be thought of as the level of accessibility, or generalized cost, faced by each user, which is represented by the destination choice logsums. Quantity is the dependent variable representing what actually was consumed given the price (or accessibility); in this case, the quantity is the number of people affected by the choice situation.

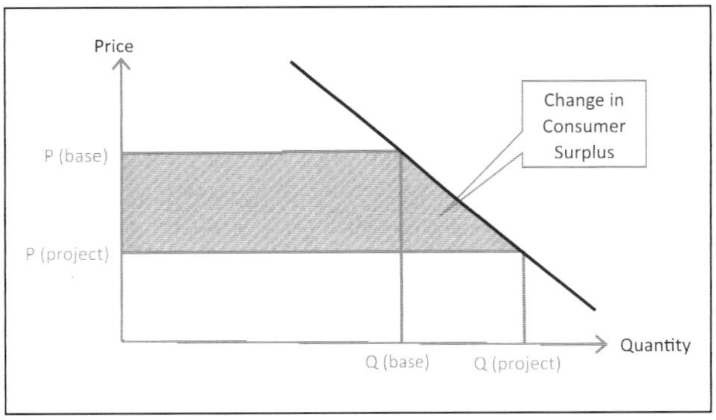

Figure 1. Change in consumer surplus.
Note: The word *base* is used as shorthand for the No Project baseline.

The rule of one-half, a common estimate of consumer surplus for transportation projects, has been shown to simply translate to the application of destination choice logsums through the following equation (*12, 13, 17*) :

$$\Delta CS = \frac{1}{2}\left(T_{\text{project}} + T_{\text{base}}\right)\left(L_{\text{project}} - L_{\text{base}}\right)$$

Here, T_{base} and $T_{project}$ are the persons traveling in the base and project model runs, L_{base} and $L_{project}$ are the destination choice logsums transformed to person hours in the base and project model runs.

In Travel Model One, destination choice logsums were calculated for 24 categories of household/user demographics:

- Four income categories: very low, low, medium, and high;
- Three auto/worker categories: zero autos, fewer household autos than household workers, and household autos not fewer than household workers; and
- Two trip purpose categories: mandatory (work or school) and non-mandatory (all other trips).

Consumer surplus was calculated for a representative person of each category residing in each travel analysis zone (TAZ) using the corresponding destination and mode choice models. Note that these market segment-based destination choice (accessibility) logsums included all possible destinations, rather than a sample, which is typically used for the destination choice model at the person level. Total consumer surplus was the sum of the consumer surplus weighted by the people represented, which are the people in the relevant category of household/user demographics living in the given TAZ. Note that the auto/worker category is not fixed for a household between scenarios because it is the result of an auto ownership model. For mandatory purpose categories, workers and students were included in the estimated quantity of people affected. For non-mandatory purpose categories, the methodology assumes that all persons are affected. As explained in RSG's report (*17*), the rule of half benefits are more straightforward to calculate using destination choice (accessibility) logsums, compared with tour-based measures, because accessibilities are calculated for every market segment in the synthetic population and includes all possible tour destinations (rather than a sample).

The result of destination choice logsum differences is measured in its own unit, *utils*. Utils can be converted to minutes using the coefficient that relates utils to minutes (*13*). For the purpose of evaluating consumer surplus, utils are converted to time and every person has the same value of time, regardless of income. In this way, the method values changes in utility at the same level across income and auto/worker categories. This is also consistent with guidance from FTA with respect to their implementation of the Summit tool (*14*, *15*) . The value of time is assumed to be one-half of the median wage rate for the San Francisco Bay Area following guidance from the California Department of Transportation on value-of-time assumptions for transport projects (*18*).

The final step for implementing destination choice logsums in the benefit–cost methodology was to combine these user benefits with the full suite of benefits attributable to a project. From Table 1, the following user benefits were replaced by the new destination choice logsum term:

- Passenger vehicle travel time and operating cost
- Transit in-vehicle travel time
- Transit out-of-vehicle travel time
- Parking cost
- Walk/bike travel time.

Note that any metric listed in Table 1 that is not included in the destination choice logsum metric was calculated as before, using realized output metrics like VMT and speeds by facility type. Additionally, as destination choice logsums were only calculated for households in the region, travel time and cost savings for truck and inter-regional trips were still calculated based on travel times and travel costs. Destination choice logsums were also only evaluated for short-term decisions such as mode and trip destination. Finally, economic transfers between users and government were considered to be neither costs nor benefits, and were deducted from the logsum-based cost/benefit tally.

Pros of Destination Choice Logsum Application

Through application of the destination choice logsums for many different investment types, the authors developed a clear sense of pros and cons for this methodology. The pros included consistency with user values in the travel model and ability to visualize results on maps, which proved an invaluable debugging process and provided context for communicating results. This section describes these two advantages of the application.

Pro No. 1: Consistency with User Values in the Travel Model

One of the clearest advantages of the new methodology was increased consistency with behavioral assumptions in the travel model. Projects that improved accessibility had positive consumer surplus estimates whereas projects that decreased accessibility (or were small enough to be within model noise) had negative consumer surplus estimates. Additionally, estimates of consumer surplus resulted in a higher economic value for increases in accessibility. Illustrating this point, Figure 2 shows the total benefit, after monetization, for the evaluated projects using the destination choice logsum method as compared to the realized travel time and cost method (the method of the 2013 Plan). Although the scale is similar, the destination choice logsum method tends to show greater total benefits, particularly for transit projects. This is intuitive for two reasons: 1) more benefits are captured in the destination choice logsum method because a traveler's utility includes more components and trade-offs than travel time and travel cost, and 2) long-term benefits are captured in relation to increased accessibility to many new destinations via transit. For example, individuals living near a new heavy rail station now have walk-to-rail access to many more destinations in the Bay Area where they previously had none.

Table 2 pulls out the benefit results of two projects from Figure 2. One project extends the region's heavy rail service by 6.5 mi from the proposed Berryessa Station in north San Jose through downtown San Jose to the city of Santa Clara, and the other project proposes adding a new carpool lane to a freeway connecting to Silicon Valley. Compared with the baseline (No Project) scenario, the heavy rail extension adds 32,000 new regional transit trips and increases regional travel times by approximately 10,000 person-hours. The carpool lane project reduces regional travel times by approximately 4,000 person-hours and results in low levels of mode-

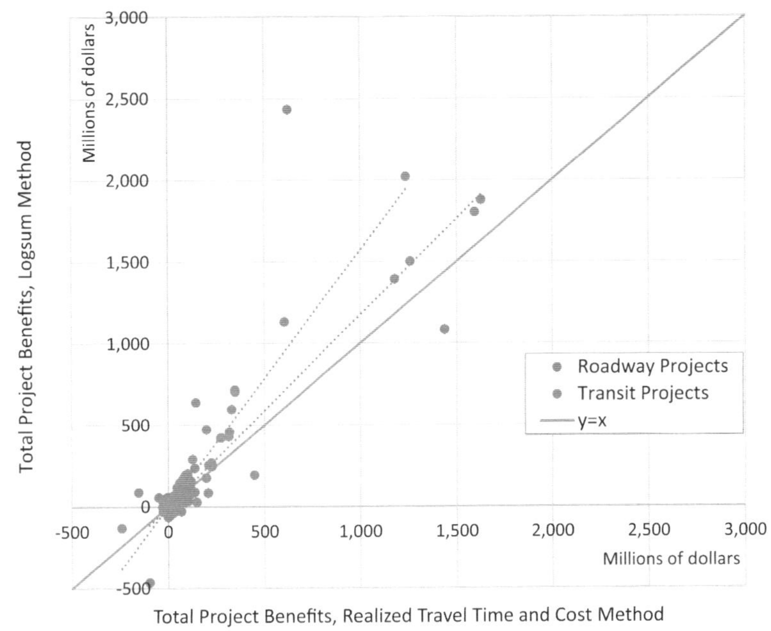

Figure 2. Logsum-based benefit vs. realized travel time and cost-based benefit for evaluated projects.
Note: Results for the realized travel time and cost method reflect an out-of-vehicle travel time adjustment; results for both methods include non-user benefits like safety and health.

Table 2. Comparison of Total Travel Time and Benefits from Both Methods Relative to a No Project Scenario

Project	Change in regional travel times	Total benefits, realized travel time and cost method	Total benefits, logsum method
Transit: Suburban heavy rail extension	9,800	US$199.8	US$472.0
Automobile: Suburban carpool lane	−3,700	US$33.0	US$63.4

Note: Values represent the change in metric relative to a baseline No Project run: positive means larger than the baseline, negative means lower than the baseline No Project; time metrics are in person-hours; benefits are in millions of dollars, normalized to year 2017, and reflect all user and non-user benefits.

switching. After the adjustment in OVTT described earlier, the estimate of total benefits, including non-user benefits like safety, for the realized method is much lower than the logsum-based method for both projects, suggesting benefits related to access and diversity in travel choices are not fully captured in the realized travel time and cost method.

Pro No. 2: Spatial Visualization of Consumer Surplus Results

A second advantage of using the destination choice logsum method was the ability to visualize the spatial distribution of consumer surplus. Figure 3 presents a map of mandatory destination choice logsum differences for the heavy rail extension presented in Table 2. The dark blue areas show where residents benefit the most from the new access to jobs along the transit system, especially to jobs in downtown San Francisco and downtown Oakland. Somewhat surprisingly, the blue areas to the north of the extension reveal an additional group of residents benefiting from access to jobs that are in the southern part of the region.

Mapping changes in consumer surplus by TAZ also became a valuable debugging tool. After each model run, the authors reviewed the consumer surplus maps and assessed whether the results reasonably corresponded to the project. If changes in destination choice logsums resulted from a coding error, the project was rerun with the corrected network, significantly increasing the accuracy of the assessment. This was required for several projects so the authors created an automated process for generating consumer surplus maps.

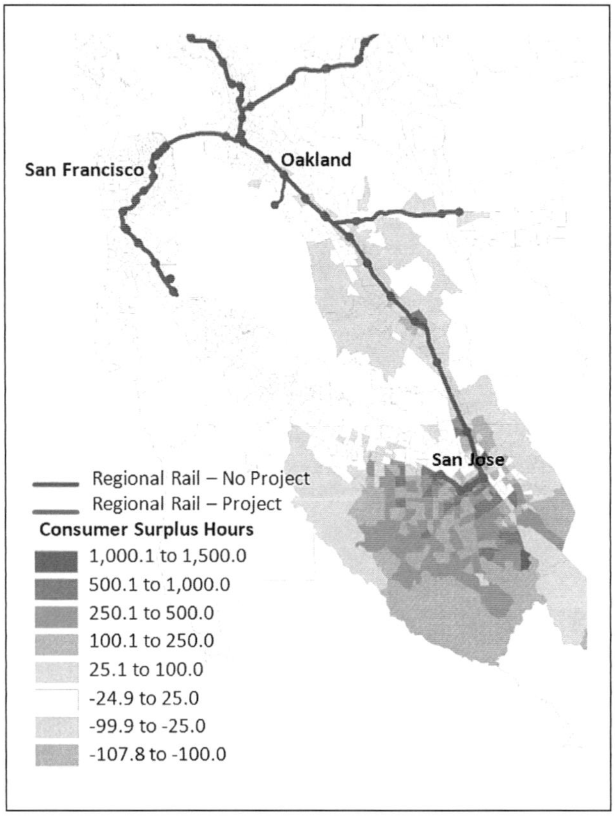

Figure 3. Consumer surplus for suburban heavy rail extension (for mandatory trip purposes).

Cons of Logsum Application

Generating maps also revealed anomalies that were not due to network problems. The authors discovered two of the cons of the logsum method from this mapping process related to structural issues with path finding in the travel model: (1) capping access travel times and (2) revealing a limited number of possible options in the estimate of mode choice. This section presents adjustments required for ultimate implementation of the logsum method. In addition to these structural issues, the authors identified a third con related to communicating the more complex estimate of consumer surplus compared with estimates of travel time and cost savings.

Con No. 1: Cliff Effects

The first con of the application relates to a limitation in Travel Model One's transit mode path finder. Figure 4 presents consumer surplus maps for a light rail extension in San Jose. The blue area in the lower right corner of the maps reflects the correct result that residents should see: an increase in access that should be relatively localized. However, the large red area in the map on the left of the figure was not intuitive.

In this example, the authors discovered that the anomaly was due to a constraint meant to limit the number of park-and-ride options considered for travelers. Travel Model One's transit mode path finder was configured to only show a park-and-ride option if the park-and-ride lot

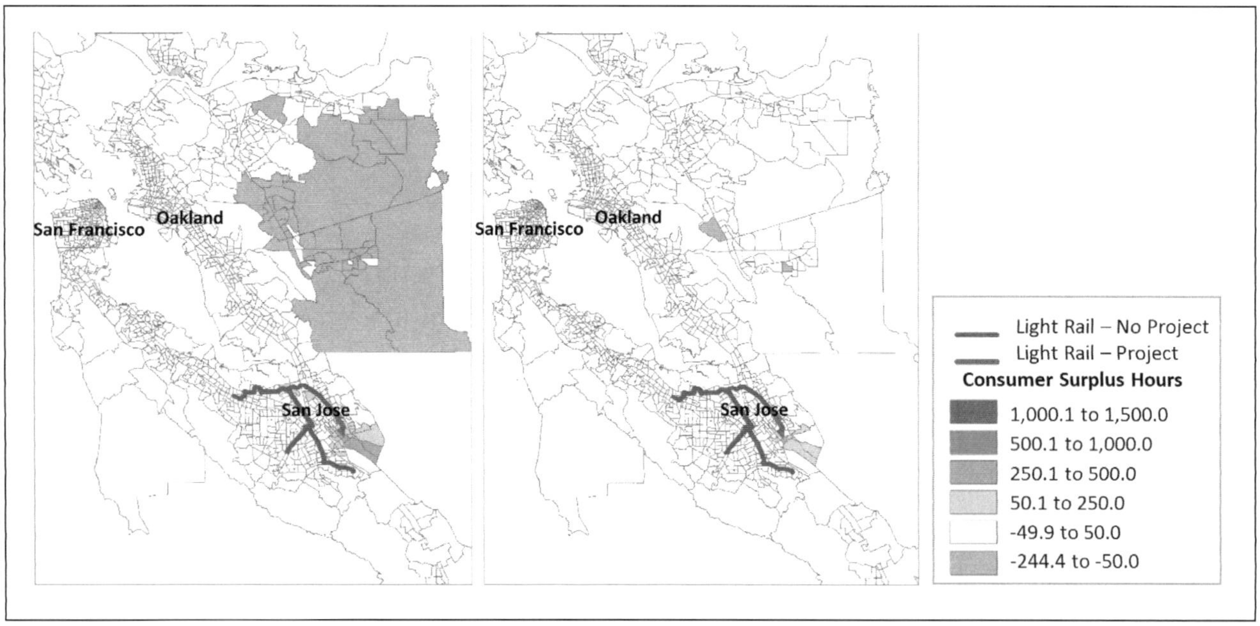

Figure 4. Consumer surplus map with cliff effects (left) and with cliff effect mitigation (right).

was 40 min away or closer to a person's origin. Capping at 40 min is typically used to constrain the number of park-and-ride lots considered for travel, which reduces model run time, and because survey data indicates that few people would consider a park-and-ride option further than 40 min away from their home. However, this "cliff effect" has unintended consequences: the stochastic nature of the model creates some variation in travel speeds on links all over the network, so the model may estimate that travel time to a park-and-ride lot is 39.9 min in the baseline and 40.1 min in the Project scenario, causing the drive-to-transit mode to become unavailable in the Project scenario. This change in accessibility is minimal, because the drive-to-transit mode was not particularly attractive in the baseline, but it has a non-trivial effect on benefit estimation when the small change is multiplied over all the residents of the areas with park-and-ride options around 40 min away from home locations.

To handle these cliff effects, the authors developed a method for cliff effect mitigation. The mitigation identified whenever the magnitude of the destination choice logsum difference was small compared with the maximum destination choice logsum difference, and concluded this difference was likely due to a cliff effect. These small destination choice logsum differences were subsequently dampened with a smooth function before being multiplied by the population affected. In this way, the limitation of the path-finding algorithm was minimized and the appropriate changes in destination choice logsums were incorporated in the estimate of consumer surplus. The result of this mitigation is the map on the right in Figure 4.

Con No. 2: Inconsistent Path Finding and Mode Choice Set Construction

A second issue was discovered related to the path-finding algorithm of Travel Model One. Consider a mode choice model with separate bus and rail alternatives and a corridor in which bus and rail compete. If the project dramatically improves bus service, the rail route may not prevail during the travel model's path-finding process when searching for a rail-eligible path, resulting in the rail alternative being dropped as a viable alternative. Depending on the size of the relative alternative-specific constants, this may result in illogical user benefit decreases. Although the number of instances in which two transit modes closely compete in the same corridor is rare, the issue arose a few times when the authors examined nonintuitive user disbenefits. For example, one project that increased San Francisco local bus frequencies (and did not decrease or otherwise affect rail service) resulted in a negative consumer surplus for a handful of TAZs where both rail and bus operate. In order to mitigate for this issue, the authors included the ability to zero out logsum differences for specific TAZs that had invalid results. Zeroing out these logsum differences generally had a small effect on overall benefit estimation and was applied only when warranted.

Con No. 3: What is a Logsum? Communicating Results

The last con involves communicating consumer surplus in logsum terms to policy makers and partner agencies. Several of the authors of previous papers have noted with surprise that logsums have not been more widely used in practical applications. De Jong et al. (*12*) provide one possible explanation: the more conventional approach of calculating realized travel time and cost differences is "more transparent" and "therefore easier to explain to non-experts." This may be a significant barrier to wider use of destination choice logsums in project appraisal. Through presentation of results, the authors struggled with describing consumer surplus and logsums in simple terms, particularly because travel time savings could not be reported separately. The authors presented the new benefit–cost methodology as valuing accessibility by capturing travel time and cost changes across all possible destinations, which is more consistent with planning objectives of increasing alternative mode share and reducing greenhouse gas emissions than simply reducing travel times and costs. The authors believe that this direct connection to accessibility increased the level of comfort with the more complex performance measure.

Conclusion

This paper presents several pros and cons for using the destination choice logsum methodology in a practical application of benefit estimation. The main advantages of using a logsum-based approach to assess user benefits were: 1) consistency with utility estimation in the travel model (e.g., able to reflect benefits of introducing a new mode) and 2) the ability to visualize a spatial representation of user benefits, which aided in communicating the distributional benefits of transport projects as well as in assessing quality control in network coding.

The authors did need to solve a few challenges in the application of the logsum-based method. The "cliff effects" inherent to model path-finding parameters created some unintuitive results, requiring the development of a cliff effect mitigation. Additionally, although modelers seek to avoid inconsistencies between the model's path finder and the model's mode choice model, some inconsistencies are difficult to avoid. Although these inconsistencies had minimal effects, when they did arise, the authors discounted the consumer surplus values in the affected TAZs. Finally, the logsum methodology may be more difficult to explain to non-technical audiences, even as the methodology is becoming better known among modeling practitioners.

Despite the cons of the methodology, the authors feel that the advantages of having a measure that is more consistent with the behavioral values reflected in the activity-based travel model, along with the value of having an intuitive visualization of the consumer surplus for each project, outweigh the cons. Above all, quantifying transportation benefits has been found to be an invaluable tool for prioritizing projects in the San Francisco Bay Area region. When using an activity-based travel model to forecast project benefits across a range of different project types, using destination choice logsums is an appropriate choice as it is technically a more accurate method for estimating consumer surplus. Fundamentally, it provides a more accurate estimate of the user benefits derived from transit projects when mode-switching and increases in access are large outcomes of the project. Given the difficulties in explaining logsums and consumer surplus in isolation of other benefit categories, and the adjustments that were still required with this methodology, the authors recommend estimating the conventional benefits of travel time and cost savings alongside the estimate of consumer surplus through changes in destination choice logsums to provide the full context of the benefits of transport projects.

Author Contributions

The authors confirm contribution to the paper as follows: study conception and design: DO, DV, LZ; data collection: KV; analysis and interpretation of the results: LZ, KV, DO, DV; draft manuscript preparation: KV, LZ. All authors reviewed the results and approved the final version of the manuscript.

References

1. Vautin, D., L. Klein, S. Co, K. Jeannotte, and D. Sallman. Leveraging Project-Level and Scenario-Level Performance Assessment to Achieve Sustainability Goals of Plan Bay Area. Presented at 92nd Annual Meeting of the Transportation Research Board, Washington, D.C., 2013.
2. Plan Bay Area 2040. Metropolitan Transporation Commission and Association of Bay Area Governments, Oakland, CA, 2017. http://2040.planbayarea.org/
3. *Transportation 2035: Change in Motion.* Metropolitan Transportation Commission and Association of Bay Area Governments, Oakland, CA, 2009. http://mtc.ca.gov/our-work/plans-projects/plan-bay-area-2040/transportation-2035. Accessed July 21, 2017.
4. *Plan Bay Area: Strategy for a Sustainable Region.* Metropolitan Transportation Commission and Association of Bay Area Governments, Oakland, CA, 2013. http://files.mtc.ca.gov/pdf/Plan_Bay_Area_FINAL/Plan_Bay_Area.pdf. Accessed July 21, 2017.
5. Paterson, L., and D. Vautin. *Integrating Local Street Maintenance Projects into a Performance-Based User-Oriented Project Evaluation Framework.* 2015. https://trid.trb.org/view.aspx?id = 1336990. Accessed July 29, 2017
6. Paterson, L., and D. Vautin. Evaluating the Regional Benefit/Cost Ratio for Transit State of Good Repair Investments. *Journal of Public Transportation*, Vol. 18, No. 3, 2015, pp. 15–28. http://dx.doi.org/10.5038/2375-0901.18.3.2.
7. *Plan Bay Area 2040 Final Supplemental Report: Performance Assessment Report.* Metropolitan Transportation Commission and Association of Bay Area Governments, San Francisco, CA, 2017. http://2040.planbayarea.org/sites/default/files/2017-07/Performance%20Assessment%20Report_PBA2040_7-2017_0.pdf. Accessed July 21, 2017.
8. *COst Benefits Results Analyzer (COBRA) Scripts.* Metropolitan Transportation Commission, San Francisco, CA, 2017. https://github.com/MetropolitanTransportationCommission/travel-model-one/tree/master/utilities/PBA40/metrics. Accessed July 21, 2017.
9. *Plan Bay Area: Final Performance Assessment Report.* Metropolitan Transportation Commission, San Francisco, Calif., 2013. http://www.planbayarea.org/sites/default/files/pdf/final_supplemental_reports/FINAL_PBA_Perfor mance_Assessment_Report.pdf. Accessed July 21, 2017.
10. Williams, H. C. W. L. On the Formation of Travel Demand Models and Economic Evaluation Measures of User Benefit. *Environment and Planning*, Vol. 9, No. 3, 1977, pp. 285–344.
11. Small, K. A., and H. S. Rosen. Applied Welfare Economics with Discrete Choice Models. *Econometrica*, Vol. 49, No. 1, 1981, pp. 105–130.
12. De Jong, G., A. Daly, M. Pieters, and T. van der Hoorn. The Logsum as an Evaluation Measure: Review of the Literature and New Results. *Transportation Research Part A: Policy and Practice*, Vol. 41, No. 9, 2007, pp. 874–889.
13. Kohli, S., and A. Daly. The Use of Logsums in Welfare Estimation: Application in PRISM. Presented at the European Transport Conference, Strasbourg, France, 2006.
14. Federal Transit Administration. Presented at the Workshop on Travel Forecasting for New Starts Proposals, Minneapolis, MN, 2006.
15. Freedman, J., J. Castiglione, and B. Charlton. Analysis of New Starts Project by Using Tour-Based Model of San Francisco, California. *Transportation Research Record: Journal of the Transportation Research Board*, 2005. 1981: 24–33.
16. San Diego Forward. *Evaluating the Performance of the Transportation Network.* San Diego Association of Governments, 2015. http://www.sdforward.com/pdfs/Final_PDFs/AppendixN.pdf.
17. Resource Systems Group. *Benefit Cost Analysis Using Activity-Based Models.* Draft Final Report Prepared for the Federal Highway Administration, White River Junction, VT, 2016.
18. *California Life-Cycle Benefit/Cost Analysis Model (Cal-B/C) Technical Supplement to User's Guide.* California Department of Transportation, 1999. http://www.dot.ca.gov/hq/tpp/offices/eab/benefit_files/tech_supp.pdf. Accessed July 21, 2017.

The Standing Committee on Transportation Planning Applications (ADB50) peer-reviewed this paper (18-05794).

Article

TRR
JOURNAL OF THE TRANSPORTATION RESEARCH BOARD

Transportation Research Record
2018, Vol. 2672(46) 73–82

DOI: 10.1177/0361198118798298
journals.sagepub.com/home/trr
SAGE

Using Google's Passive Data and Machine Learning for Origin-Destination Demand Estimation

Bhargava Sana[1], Joe Castiglione[1], Drew Cooper[1], and Dan Tischler[1]

Abstract

Origin-destination (OD) data collection methods are steadily attempting to move from conventional survey techniques (roadside interview, license plate, etc.) toward using passively collected big data sources such as those based on global positioning system (GPS) and cell phone call detail records (CDR). In this study, a new passive data source, Google's Aggregated and Anonymized Trips (AAT), was used to derive hourly OD demand matrices for the San Francisco Bay Area. Since the AAT dataset contains relative flows or weights as opposed to absolute trips, machine learning techniques were applied to convert them with the help of observed OD flows from expanded household travel survey. Several machine learning models were trained to perform quite well for both training and test data. However, it was found that the multi-layer perceptron (MLP), a neural networks approach, resulted in the best performing model for the conversion. Additionally, all models were used for predictions in a hypothetical application context where input AAT data were scaled by different growth factors. This exercise showed that, even though the trip predictions of all models were close to each other initially, they varied widely for different magnitudes of OD markets and growth factors.

Origin-destination (OD) data collection methods are steadily attempting to move from conventional survey techniques (roadside interview, license plate, etc.) toward using passively collected big data sources such as those based on global positioning system (GPS) and cell phone call detail records (CDR). These data sources usually have the advantage of both higher spatial coverage and higher sample sizes (*1*). In addition, most of these data are collected continuously and for longer periods of time. However, such data usually lack a high level of detail on information such as socio-demographics and trip purpose. Past studies have identified the need to understand how and to what extent these new data sources can be leveraged to improve transportation planning (*2, 3*).

A recent study by the authors (*4*) used a new dataset from Google which is based on location reports generated from mobile device users who opted to store them on Google servers. The dataset, called Aggregated and Anonymized Trips (AAT) data, was originally obtained for estimating hourly facility-specific OD matrices for the Freeway Corridor Management Study (FCMS) in San Francisco. As mentioned in the study (*4*), AAT data contain relative flows or weights instead of actual trip counts. Google does not provide absolute counts because the location reports only account for a sample of travelers. Consequently, linear regression models were estimated to convert these weights to trips using Caltrans's Performance Measurement System (PeMS), a traffic counts database. However, in addition to including facility-specific hourly OD matrices of relative weights, the dataset also contains overall (i.e., non-facility-specific) hourly OD matrices of relative weights for the nine-county San Francisco Bay area. It was determined that the region's most recent household travel survey dataset from the California Household Travel Survey (CHTS) 2012 could be used to develop models for converting the hourly OD matrices of weights in the AAT dataset to hourly OD matrices of person trips. Given that this modeling effort was focused more on establishing a relationship between weights in AAT data and observed trips rather than explaining traveler behavior choices, the authors expanded the set of estimation techniques beyond the linear regression model to evaluate machine learning (ML) techniques also. ML techniques are gaining popularity in the transportation

[1]San Francisco County Transportation Authority, San Francisco, CA

Corresponding Author:
Address correspondence to Bhargava Sana: bhargava.sana@sfcta.org

domain (*5–13*). Previous studies have found that ML models are better at capturing non-linear relationships in the data compared with statistical models and are also less restrictive in terms of the assumptions regarding error structures, making them arguably more appropriate for this effort.

The main objective of this study was to use Google AAT data and observed OD demand from CHTS and model a relationship between anonymized weights and actual trips. The model could then be used to predict dynamic OD person trip matrices using current and future versions of AAT data. A secondary objective was to train and evaluate several different ML models using AAT data and add to the growing literature examining the applicability of these models in the travel modeling domain. The various models were compared based on their prediction accuracy for a subset of data that was set aside for testing. In addition, the sensitivity and predictive nature of the models was evaluated using a hypothetical application context in which existing AAT weights are scaled to represent the effect of a new investment or policy.

This paper is organized as follows. The next section provides details about relevant literature on both OD studies using passively collected big data sources and the use of ML techniques. The following section briefly introduces the data used and the methods section highlights the specific ML techniques applied in this effort. The results section presents a comparison of the performance and accuracy of different models and the final section provides conclusions and discussion.

Literature Review

An increasing number of studies are being published each year that evaluate the use of emerging passively collected big data sources for estimating OD demand for a region. These data sources are primarily based on aggregated and anonymized cell phone CDR or GPS traces (*14–20*). However, it must be noted that data from smart fare cards have also been used to derive transit demand (*21*) and social networking data have been used to estimate non-commute trips (*22*).

Recognizing the fact that data obtained from these big data sources are still samples from the whole population, the datasets need to be processed and/or expanded to obtain an estimate of the full OD demand matrix. The expansion factors are usually a function of the penetration rates of various technologies which in turn are calculated using some source of observed data (surveys, traffic counts, etc.). A recent study in Texas (*20*) compared Bluetooth, cell phone and GPS data sources for obtaining external–external (E-E) OD demand. The study observed that different data sources could have different biases and constraints. For example, GPS data could be more biased toward commercial vehicle trips and cell phone data might have relatively lower spatial and temporal resolutions. The study also found that the total demand estimated from cell phone data was significantly lower than that estimated from GPS and Bluetooth sources. GPS and Bluetooth data were expanded using traffic counts at external stations whereas cell phone data were expanded using census tract population data. A few other studies have also used population data from national census to expand cell phone OD data and obtain OD matrices at the regional level (*16–18*). In other cases, the initial OD demand from sample has been assigned and iteratively adjusted to match observed roadway link volumes closely (*14*, *15*).

The current effort posed a different challenge because Google did not provide any absolute (sample) numbers in the dataset. As mentioned in the previous section, relative flows or weights were given for each OD by hour instead. Additional details regarding the dataset are provided in the next section. Also mentioned in previous section was the motivation to convert these weights into trips by applying ML techniques. A recent household travel survey for the region, CHTS 2012, was identified to provide the target trips for training the ML models. It must be noted that the CHTS target trips are themselves estimates expanded from a sample using regional population data.

ML techniques were being applied in the field of travel choice modeling from as early as the 1990s (*5*). Since then, different studies have applied a variety of advanced ML models (neural networks, support vector machine, random forests, etc.) to predict a range of travel dimensions (demand, travel mode, purpose, time of day, etc.). Mohammadian and Miller (*5*) compared artificial neural networks (ANN) and nested logit models for predicting household automobile choice (both vehicle class and vintage) and found that while both techniques resulted in strong models, ANN showed higher prediction accuracy. A more recent comparison between ANN and copula-based joint statistical models (*12*) for trip mode and departure time concluded that ANN was not only faster and easier to implement but also had a slightly better prediction performance. An ANN model was also applied to model activity scheduling in Quebec, Canada (*6*). Several other studies have applied ML techniques to identify travel mode. Pirra and Diana (*13*) used the support vector machine (SVM) method to predict tour mode in New York State using National Household Travel Survey (NHTS) data, while Elhenawy and Rakha (*10*) used a combination of random forest and hidden Markov models to predict trip mode from smartphone sensor data. Random forest has also been applied to classify trip purpose using GPS and accelerometer data (*8*). Hidden

Markov models and cell phone CDR data have been used to generate full-day activity patterns (*1*). A recent study used NHTS and Census Block Group data and applied support vector regression (SVR) to predict weekly non-motorized trips on rural roadways (*11*).

Overall, past studies have found that, even though ML techniques can be tuned to predict better than statistical models, it is difficult to interpret the significance and elasticity effects of variables in these models. Therefore, they appear to be more appropriate if the goal is accurate prediction of a base condition rather than policy evaluation and analysis of alternatives. The current study aims to add to this growing body of literature and provide further insights into the applicability of ML techniques to processing a big data source in the context of estimating OD demand.

Data Description

The OD dataset from Google was obtained through the Better Cities program (*23*) for the San Francisco FCMS. The data were derived from location reports of mobile device users who had chosen to store their information on Google servers. The location reports of signed-in Google users are generated using a range of sensor types including GPS, cell towers, and Wi-Fi detectors. The raw data undergoes several steps of processing before being made available. Identifying the users that are moving, cleaning location data, and applying differential privacy filters to protect user anonymity are a few of those steps. In the final steps, after the spatial and temporal aggregation of OD data, the absolute trips are converted to relative flows or weights. This is done by assigning the record with the highest number of absolute trips in the full dataset to one and by calculating the relative weights of all the other records as ratios of respective absolute trips to the highest absolute trips. Therefore, only one record (for a particular OD, day, and hour) was assigned a relative weight of one. Since Google's location reports are generated from personal mobile devices, AAT flows most likely represent person trips/flows.

An 85-district system covering the nine-county San Francisco Bay Area was defined (Figure 1). For San Francisco and San Mateo counties, the districts were defined based on TAZs but for the remaining counties, district boundaries were the same as county boundaries. OD data aggregated to this district system and one-hour time intervals were obtained for each day in a six-month period in 2015 (April–June and September–November). More details about this dataset can be found in a recent study by the authors (*4*).

CHTS 2012 was conducted with goal of updating the statewide database on household travel behavior. The San Francisco Bay Area sample consisted of roughly 8,000 households, 20,000 persons, and 70,000 trips. Weights for expanding survey records were developed

Figure 1. San Francisco Bay Area districts.

Table 1. Weekday Descriptive Summary of Google AAT and CHTS Data

Data	Count	Minimum	Maximum	Mean	SD	Sum
Google AAT hourly OD	145,758	0.0002	0.9257	0.0007	0.0118	99
CHTS hourly OD	9,969	52	404,951	2,311	18,256	23,047,002
Google AAT daily OD (aggregated)	7,225	0.0028	13.3745	0.0137	0.2292	99
CHTS daily OD (aggregated)	2,996	61	4,676,065	7,692	129,297	23,047,002

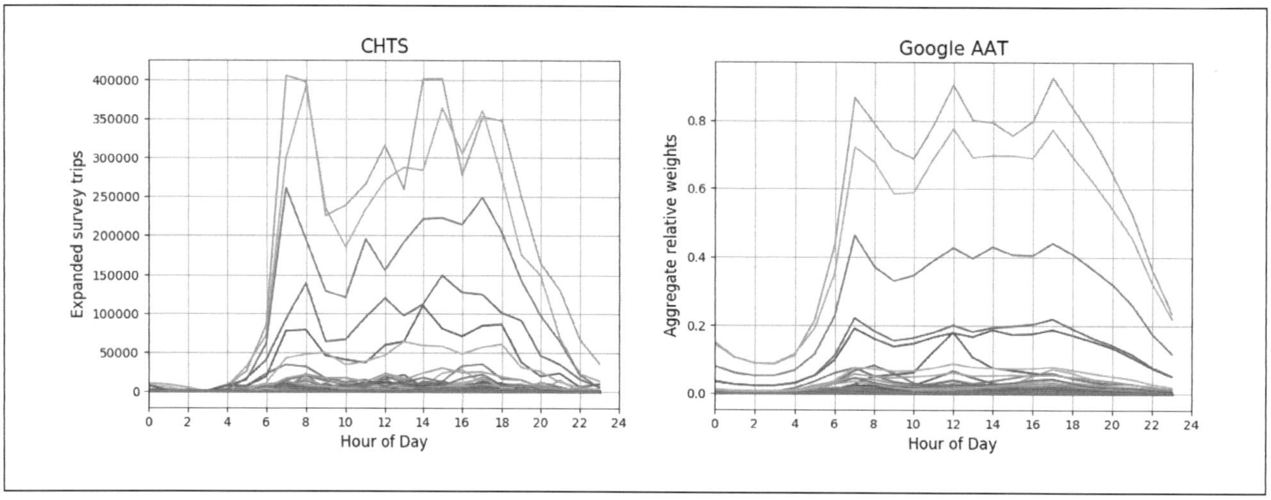

Figure 2. Time of day distributions of CHTS and Google AAT OD flow data (weekday).

using regional population data. Considering that this dataset was the latest observed source for regional OD flows, the expanded survey records were used to obtain target hourly OD person trip demand values for the conversion of Google weights.

Origin and destination trips ends of CHTS trips were mapped to the 85-district system and weighted sums were calculated to obtain hourly OD flows. The raw AAT dataset was further processed to obtain an average hourly weight for each OD across the six-month period. Only weekday records from both Google AAT and CHTS datasets were used. Table 1 shows a descriptive summary of the weights in two data sources. As expected, Google data had a much larger spatio-temporal coverage compared with CHTS. At a daily level, the big data source contained flows for all the OD pairs ($85 \times 85 = 7{,}225$) whereas CHTS had information for less than half that number (2,996). A broad range of average weekday hourly OD weights was observed in the AAT data (0.0002 to 0.9). The total daily trips for the region indicated by the expanded survey is just over 23 million. Figure 2 shows the plots of expanded OD flows for CHTS and flows from Google AAT weights by time of day. Each line in the plot represents the average weekday distribution of demand by hour of day for a particular OD market. At a high level, distributions of big OD markets corresponding to data in CHTS matched with those in the Google dataset. However, for smaller OD markets, larger differences in distributions were observed. There appears to be more noise in the CHTS flows. The plots also show that the relationship between observed flows and flows from Google AAT data varies both by time of day and OD market.

Methods

ML techniques broadly fall into two categories: supervised and unsupervised. In supervised learning, the input data contain the target variable that needs to be predicted, whereas in unsupervised learning there is no prior knowledge about the outcomes and the algorithms are expected to identify the underlying structure and associations. This study focuses only on supervised ML techniques.

The general approach in supervised learning methods is to split the input data into "train" and "test" subsets (usually a random 75%–25% split). The algorithm will attempt to "learn" the structure by "fitting" the input features (independent variables) and target (or dependent) variable in the training dataset. This could be a process as simple as minimizing squared errors between predictions and true target values. The next step involves using the "trained" model to make predictions on the test dataset. The prediction accuracy of the model on the test dataset would indicate the level of generalization that the training was able to achieve. There is always a risk of overfitting which needs to be minimized. The training and application of all ML techniques was done using the "scikit-learn" Python library (*24*). This library provides a very convenient and easy-to-use application programming interface (API) to apply ML techniques. In this effort, seven different ML techniques were explored and are described next. It must be noted at the outset that all the techniques are succinctly described here and more detailed treatments can be found in other previous studies (*5–13*).

Ordinary Least Squares (OLS), Ridge, and Lasso Regression

In generalized linear regression models, the target/dependent variable (y) is modeled as a linear function of a set of (m) independent variables (x) and their respective coefficients/parameters (b) as shown in Equation 1. Ordinary least squares (OLS) regression minimizes the sum of residual squares between the target variable (y_{obs}) and that predicted by the linear approximation (y_{pred}) as shown in Equation 2. Here, B represents the vector of coefficients (b_0, b_1,..., b_m) and X represents the vector of independent variables (x_0, x_1,..., x_m) and $|| \, ||_2$ represents ℓ^2 norm.

$$y_{pred}(b,x) = b_0x_0 + b_1x_1 + b_2x_2 + b_3x_3 + \ldots + b_mx_m = XB \quad (1)$$

$$\min_b ||XB - y_{obs}||_2^2 \quad (2)$$

$$\min_b ||XB - y_{obs}||_2^2 + \alpha||B||_2^2 \quad (3)$$

$$\min_b \frac{1}{2n}||XB - y_{obs}||_2^2 + \alpha||B||_1 \quad (4)$$

OLS estimation relies on the independence of explanatory variables (x), failing which the coefficient estimates (b) could become very sensitive to random errors in observed target variable and could produce high variance. To prevent any of the coefficients from becoming extremely large and to improve performance in case of multi-collinearity, the objective function is modified in Ridge and Lasso regressions as shown in Equations 3 and 4 respectively. In Equation 4, n denotes the number of samples and $|| \, ||_1$ represents ℓ^1 norm. The additional parameter α is known as the regularization parameter which controls the size of coefficients. Lasso regression could drive certain coefficients to zero and is often used to reduce the number of independent variables.

k-Nearest Neighbor Regression

k-nearest neighbor or KNN regression is one of the simpler ML algorithms and it does not have any parameters to estimate. The prediction is based on the observed target values of "k" nearest neighbors in terms of a distance metric (Euclidean, Manhattan, etc.) in the multi-dimensional variable space. It could be an average, median, maximum, or any other statistic of the observed target values of the nearest neighbors. For the current study, the scikit-learn default values were used: five nearest neighbors, Euclidean distance, and weighted average for the predicted value where nearer neighbors will influence the prediction more than the farther ones.

Support Vector Regression

SVR attempts first to transform the input variables (x) to a higher dimensional hyperspace using a fixed non-linear function and then constructs the linear model in this higher dimensional space. The objective function in the linear model ignores error within a tolerance level which is also known as ε-insensitive loss function. The determination of actual non-linear function is avoided by constructing the dual of the optimization problem and use of what is called a kernel function. The most frequently used kernel is the radial basis function: exp $(-\gamma|| x_i\_x_j||^2)$ where the parameter $\gamma>0$ and x_i and x_j are two input data vectors. Providing further mathematical details is beyond the scope of this study but they can be found in other studies in the literature (*11*).

Random Forest Regression

Random forest (RF) regression involves prediction based on a set of randomized decision trees (*8*). A decision tree identifies a set of rules for the input variables that can be used to predict a target variable. In this case, a decision tree would mean learning what different ranges of Google AAT weights would result in what values of hourly OD trips. Since just using one decision tree often leads to overfitting, multiple decision trees are trained with random portions of the training dataset to form a "random forest" of the decision trees. The final prediction is an average of the individual predictions of all decision trees in the RF.

Neural Networks: Multi-Layer Perceptron (MLP)

There are many neural networks approaches but the MLP algorithm is one of the more widely used ones. This essentially maps a layer with input variables (input layer) to the target variable. In between the two, there could theoretically be unlimited "hidden" layers. Each hidden layer contains a user-specified number of nodes

Table 2. Performance Evaluation of Machine Learning Models

Model	R-sqrd test	R-sqrd train	RMSE test	RMSE train	WAPE test	WAPE train
OLS regression	0.9541	0.9778	3,549.93	2,831.72	41.22%	35.97%
Lasso regression	0.9539	0.9778	3,559.19	2,834.67	41.67%	36.19%
Ridge regression	0.9541	0.9778	3,550.75	2,831.81	41.28%	36.01%
Support vector regression	0.9438	0.9742	3,929.32	3,055.74	40.52%	33.13%
K-nearest neighbors regression	0.7897	1	7,602.81	0	62.31%	0.00%
Random forest regression	0.8071	0.9608	7,280.02	3,763.11	61.88%	24.52%
Neural networks (multi-layer perceptron)	0.9594	0.9786	3,339.80	2,780.82	39.85%	36.23%

Note: R-sqrd = coefficient of determination; 1 = sum of squared residuals/total sum of squares; RMSE = root mean squared error; WAPE = weighted absolute percent error.

or neurons and each one of these calculates a weighted sum (weights are randomly initialized at first) of outputs from the previous hidden layer and applies non-linear transformation using what is known as the activation function (Ex. Hyperbolic tangent) to produce yet another output. Similarly, the outputs generated by one hidden layer then feed into the next hidden layer until the final target variable is predicted. The error in the prediction is propagated back to adjust weights applied in different hidden layers. One advantage of neural networks is that they can learn highly non-linear models. However, there can be many local minima and different initial random initialization of weights could result in different prediction accuracies. Further details about this technique and its applications can be found in past studies (*5, 6, 12*). In the MLP algorithm applied in this study, one hidden layer of size 100 (number of neurons) was used along with a rectified linear unit ($f[x] = \max[0, x]$) activation function.

Results

There were just under 10,000 OD-hour records common to both CHTS and Google AAT datasets. A random 75%–25% split resulted in about 7,500 and 2,500 records in the training and test datasets respectively. For each ML technique described in the previous section, the same training and test datasets were used. The target variable was the hourly OD trip flow from expanded CHTS and input variables were hourly weights in Google AAT data. Parameter tuning was also performed for each of the ML models to identify the best fitting version (highest R-squared value) in each case. The tuning of various parameters (such as α in case of lasso regression and γ in case of SVR) was facilitated by functions in scikit-learn API which allows a grid search to be implemented. The models are trained repeatedly with different combinations of parameters specified by the user to determine the optimal set of parameters for each ML technique. In the next step, the trained models were applied to predict OD flows for records in test data. Finally, error statistics were calculated for both training and test data predictions. A comparison of the performance metrics of the various ML techniques is shown in Table 2.

With rigorous parameter tuning, almost all models performed quite well except KNN and RF models. As mentioned earlier, KNN is one of the simpler algorithms and it tries to "remember" the training data. Generalization is not one of its strengths. Higher prediction accuracy for training data and relatively lower prediction accuracy for test data indicates overfitting. The RF model works by combining multiple decision trees and it performs slightly better than KNN. The neural network (MLP) model was found to result in the highest R-squared and lowest RSME and weighted absolute percent error (WAPE) for both training and test datasets. However, it must be noted that MLP is one of the most complex ML techniques and has a much higher number of parameters than most other models. Given that, it is probably not very surprising that it performed the best. Figure 3 shows the predicted results for each of the models for a sample of OD markets along with aggregated regional demand. As indicated by high R-squared values (Table 2), all the models were found to be performing well both at the regional level and for large OD markets.

To understand how these models might perform in a future application context, various growth/scaling factors were applied to the Google AAT weights and used as inputs to the trained ML models. This exercise would also shed light on how sensitive the different model outputs were to the input data. To analyze the outputs, distribution of total regional predicted trips by different OD market size bins in CHTS were plotted for a few growth factors (Figure 4). Since most of the trained ML models showed good prediction accuracies in the base scenario, the distributions were close to what was indicated by CHTS (growth factor = 1). Despite this, an interesting observation is that the models produce increasingly different forecasts as the growth factor increases. For

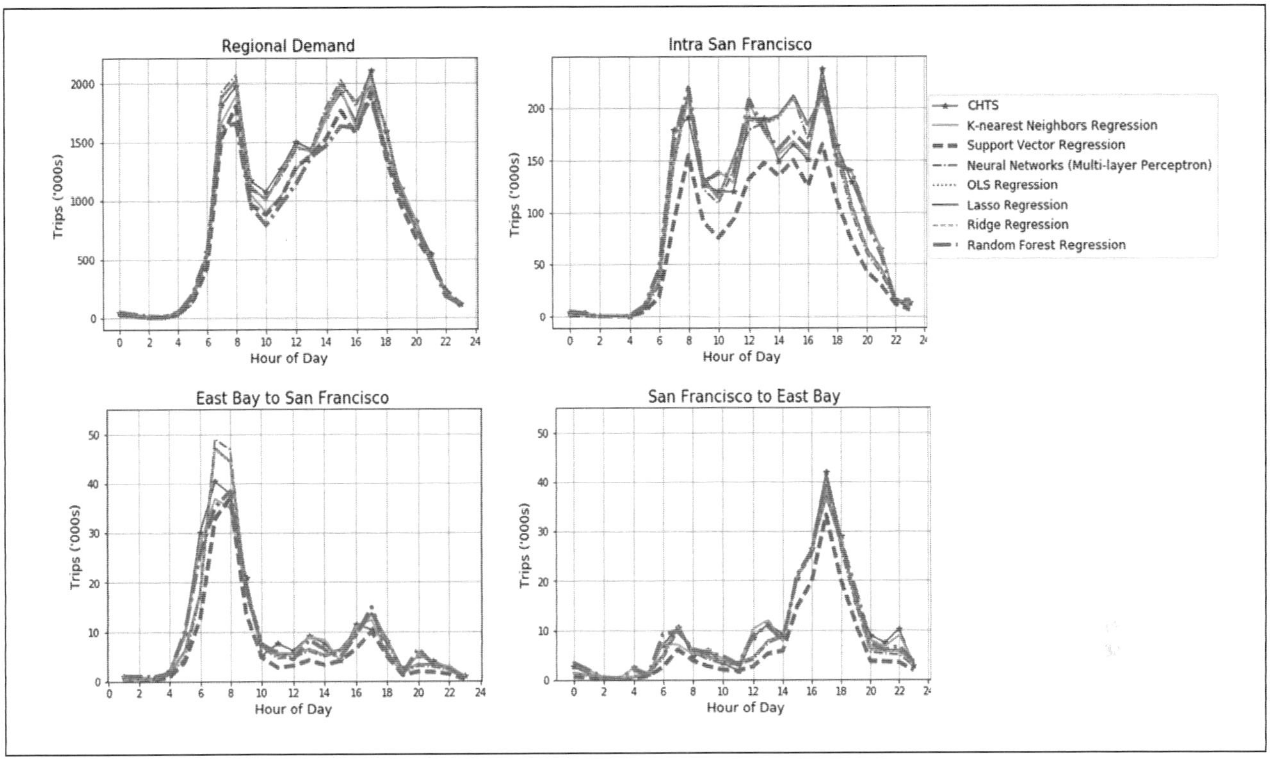

Figure 3. Comparison of ML model predictions by time of day.

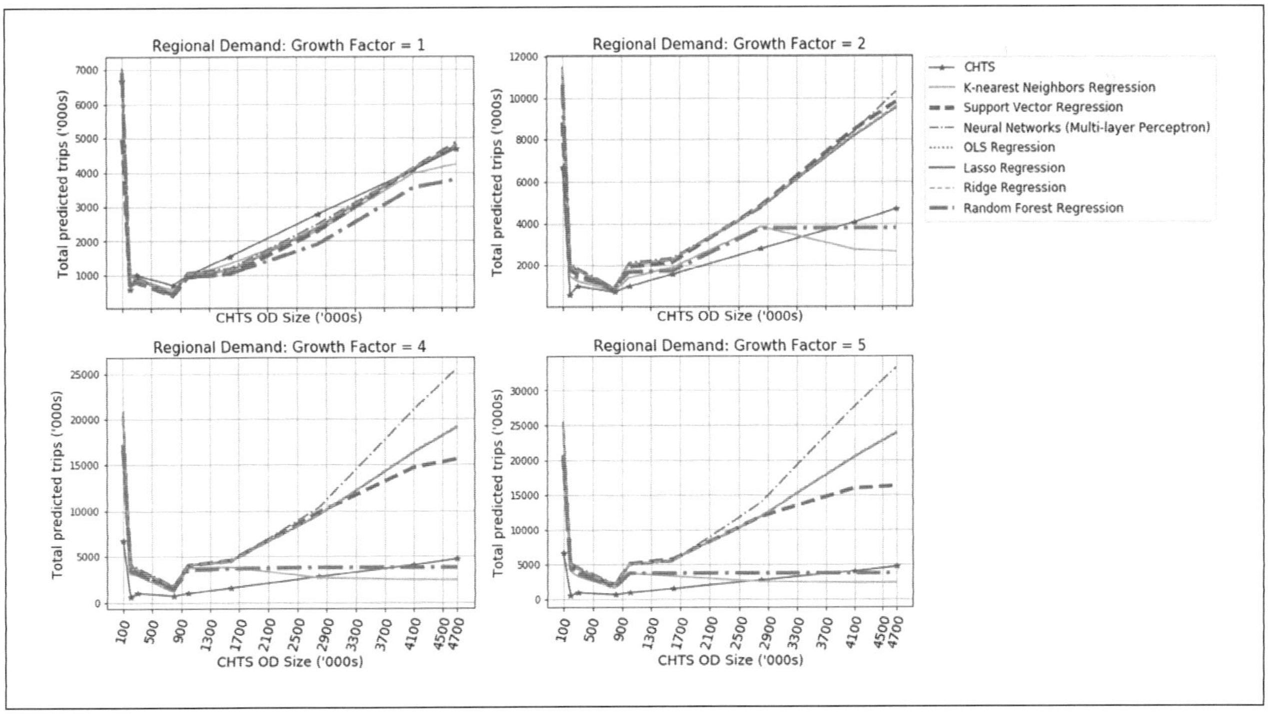

Figure 4. Distribution of total regional predicted trips by CHTS OD market size.

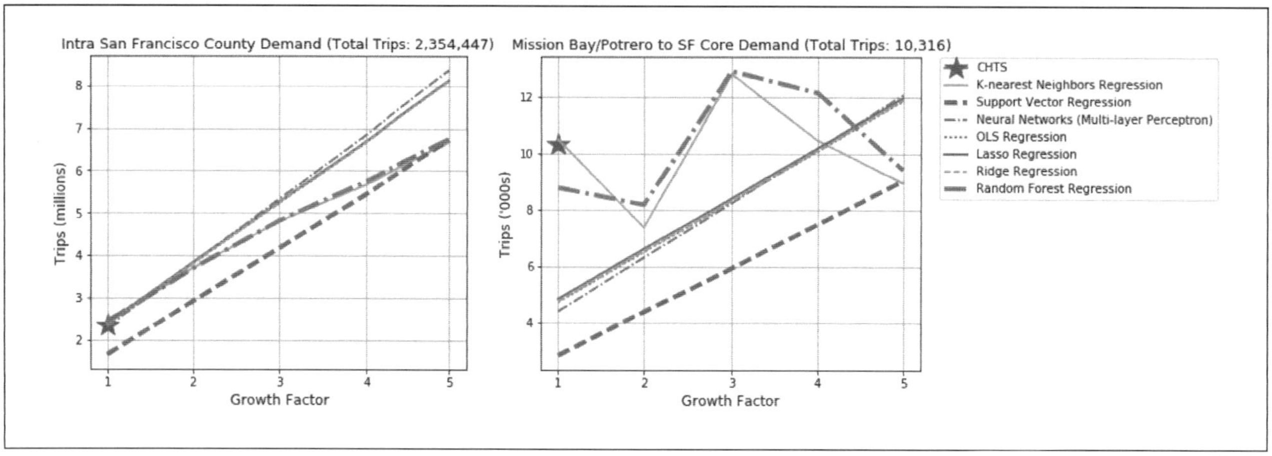

Figure 5. Comparison of ML predicted trips by growth factor and market.

example, when the growth factor is 2, KNN and RF models produce significantly lower estimates for OD markets with large numbers of trips, lower even than the original demand. When the growth factor is 4, the neural network model predicts significantly higher estimates for OD markets with large numbers of trips. The generalized linear regression models (OLS, Lasso, and Ridge), all perform similarly at different OD size levels and growth factors.

Predicted trip outputs for several OD markets and growth factors were evaluated in detail to further understand the variation. Figure 5 shows the plots of predicted trips versus growth factor for two markets: intra-San Francisco (SF) County and Mission Bay/Potrero to SF Core. For intra-SF County (whose market size is over 2 million trips per day in CHTS), the highest prediction appears to be about 30% more than the lowest prediction for all growth factors used. However, the same metric is more than 100% for the Mission Bay/Potrero to SF Core market which is a much smaller market (about 10,000 trips daily). Comparing the two plots, one could also observe that the magnitude of predicted trips by different models does not always follow the same order. For example, the highest predicted trips could be from the MLP model in one instance and the KNN model in another. All the above observations could have significant implications when the models are used in an application setting.

Conclusion

A relatively new passive OD data source from Google was used to estimate hourly person trip demand for the San Francisco Bay Area. Since this dataset from Google had only relative flows or weights for hourly OD records as opposed to absolute trip counts, seven different ML models were trained using target OD person trips from CHTS. Except KNN and RF models, all the other models could be trained to perform reasonably well for both training and test datasets as demonstrated by the high R-squared values. Interestingly, though the predictions of most ML models were close to each other initially, they varied more widely with increasing growth factors applied to the input AAT data. It appeared that markets at the lower and higher ends of the OD market size spectrum were adversely affected, which could be an artifact of the highly non-linear nature of the more advanced ML techniques. This could have significant implications when applying these models to estimate demand from future versions of this dataset which might include effects such as high growth in jobs, increased Uber/Lyft trips, or automated/connected vehicles.

Most previous studies applying ML techniques found their predictive accuracy to be better than that of statistical models. However, literature also acknowledges that ML models are much harder to interpret and explain. This study has also resulted in similar findings that although ML models can be trained to replicate some given data, the predictions and sensitivities cannot always be explained as in the case of conventional statistical models. It appears that one should be cautious before applying such models to different application contexts so that one does not end up with results that are unintuitive and can undermine confidence in the model. This raises the question of whether it is preferable to develop models with higher prediction accuracy but with relationships that are harder to explain or to settle for slightly less accuracy with higher interpretability. This could be an important question to answer before widespread application of ML techniques in the transportation modeling domain.

Another question arises from the black-box nature of most of these passive big data sources. Since these datasets are "shallow" compared with the much richer conventional survey datasets, it is harder to cross-reference them against other data sources and to expand and obtain estimates. Should the process of collecting and processing these data be made more transparent so that it enables modelers to apply more appropriate techniques? Perhaps a combination of some transparency about datasets such as Google AAT and testing their robustness across different regions and contexts would make them easier to adopt and use over the longer term.

Author Contributions

The authors confirm contribution to the paper as follows: study conception and design: Bhargava Sana, Joe Castiglione, and Dan Tishcler; data collection: Bhargava Sana and Joe Castiglione; analysis and interpretation of results: Joe Castiglione, Bhargava Sana, and Drew Cooper; draft manuscript preparation: Bhargava Sana. All authors reviewed the results and approved the final version of the manuscript.

References

1. Yin, M., M. Sheehan, S. Feygin, J.-F. Paiement, and A. Pozdnoukhov. A Generative Model of Urban Activities from Cellular Data. *IEEE Transactions on Intelligent Transportation Systems*, Vol. 19, No. 6, 2018. pp. 1682–1696. In press.
2. Chen, C., M. Batty, and T. van Vuren. Special issue: Emerging, Passively Generated Datasets for Travel Behavior and Policy Analysis. *Transportation*, Vol. 42, No. 4, 2015, pp. 537–540.
3. Chen, C., J. Ma, Y. Susilo, Y. Liu, and M. Wang. The Promises of Big Data and Small Data for Travel Behavior (aka Human Mobility) Analysis. *Transportation Research Part C: Emerging Technologies*, Vol. 68, 2016, pp. 285–299.
4. Sana, B., J. Castiglione, D. Cooper, and D. Tischler. Using Google's Aggregated and Anonymized Trip Data To Support Freeway Corridor Management Planning in San Francisco. *Transportation Research Record: Journal of the Transportation Research Board*, 2017. 2643: 65–73
5. Mohammadian, A., and E. J. Miller. Nested Logit Models and Artificial Neural Networks for Predicting Household Automobile Choices: Comparison of Performance. *Transportation Research Record: Journal of the Transportation Research Board*, 2002. 1807: 92–100.
6. Doherty, S. T., and A. Mohammadian. Application of Artificial Neural Network Models to Activity Scheduling Time Horizon. *Transportation Research Record: Journal of the Transportation Research Board*, 2003. 1854: 43–49.
7. Abdulazim, T., H. Abdelgawad, K. M. Nurul Habib, and B. Abdulhai. Using Smartphones and Sensor Technologies To Automate Collection of Travel Data. *Transportation Research Record: Journal of the Transportation Research Board*, 2013. 2383: 44–52.
8. Montini, L., N. Rieser-Schüssler, A. Horni, and K. W. Axhausen. Trip Purpose Identification from GPS Tracks. *Transportation Research Record: Journal of the Transportation Research Board*, 2014. 2405: 16–23.
9. Farid, Y. Z., E. Christofa, and L. Paget-Seekins. Estimation of Short-Term Bus Travel Time by Using Low-Resolution Automated Vehicle Location Data. *Transportation Research Record: Journal of the Transportation Research Board*, 2016. 2539: 113–118.
10. Elhenawy, M., and H. A. Rakha. Random Forest-Hidden Markov Transportation Mode Recognition Model Using Smartphone Sensor Data. Presented at 96th Annual Meeting of the Transportation Research Board, Washington, D.C., 2017.
11. Das, S., K. K. Dixon, R. E. Avelar, and K. Fitzpatrick. Using Machine Learning Techniques to Estimate Non-Motorized Trips for Rural Roadways. Presented at 96th Annual Meeting of the Transportation Research Board, Washington, D.C., 2017.
12. Golshani, N., R. Shabanpour, S. M. Mahmoudifard, S. Derrible, and A. Mohammadian. Comparison of Artificial Neural Networks and Statistical Copula-Based Joint Models. Presented at 96th Annual Meeting of the Transportation Research Board, Washington, D.C., 2017.
13. Pirra, M., and M. Diana. Tour-Based Mode Choice Study through Support Vector Machine Classifiers. Presented at 96th Annual Meeting of the Transportation Research Board, Washington, D.C., 2017.
14. Friedrich, M., K. Immisch, P. Jehlicka, T. Otterstätter, and J. Schlaich. Generating Origin-Destination Matrices from Mobile Phone Trajectories. *Transportation Research Record: Journal of the Transportation Research Board*, 2010. 2196: 93–101.
15. Iqbal, M. S., C. F. Choudhury, P. Wang, and M. C. González. Development of Origin-Destination Matrices using Mobile Phone Call Data. *Transportation Research Part C: Emerging Technologies*, Vol. 40, 2014, pp. 63–74.
16. Çolak, S., L. P. Alexander, B. G. Alvim, S. R. Mehndiratta, and M. C. González. Analyzing Cell Phone Location Data for Urban Travel: Current Methods, Limitations, and Opportunities. *Transportation Research Record: Journal of the Transportation Research Board*, 2015. 2526: 126–135.
17. Toole, J. L., S. Colak, B. Sturt, L. P. Alexander, A. Evsukoff, and M. C. Gonzalez. The Path Most Traveled: Travel Demand Estimation Using Big Data Sources. *Transportation Research Part C: Emerging Technologies*, Vol. 58, 2015, pp. 162–177.
18. Alexander, L., S. Jiang, M. Murga, and M. C. Gonzalez. Origin-Destination Trips by Purpose and Time of Day Inferred from Mobile Phone Data. *Transportation Research Part C: Emerging Technologies*, Vol. 58, 2015, pp. 240–250.
19. Ge, Q., and D. Fukuda. Updating Origin-Destination Matrices with Aggregated Data of GPS Traces. *Transportation Research Part C: Emerging Technologies*, Vol. 69, 2016, pp. 291–312.
20. Hard, E. N., B. T. Chigoy, P. Songchitruksa, S. P. Farnsworth, D. W. Borchardt, and L. L. Green. Comparison of Cell, GPS, and Bluetooth Derived External O-D Data – Results from the 2014 Tyler, Texas Study. Presented at 96th Annual Meeting of the Transportation Research Board, Washington, D.C., 2017.
21. Alsger, A., B. Assemi, M. Mesbah, and L. Ferreira. Validating and Improving Public Transport Origin-Destination

Estimation Algorithm using Smart Card Fare Data. *Transportation Research Part C: Emerging Technologies*, Vol. 68, 2016, pp. 490–506.
22. Yang, F., P. J. Jin, Y. Cheng, J. Zhang, and B. Ran. Origin-Destination Estimation for Non-Commuting Trips Using Location-Based Social Networking Data. *International Journal of Sustainable Transportation*, Vol. 9, No. 8, 2015, pp. 551–564.
23. Eland, A. *Tackling Urban Mobility with Technology*. Google Europe Blog, November 2015. http://googlepolicyeurope.blogspot.com.es/2015/11/tackling-urban-mobility-with-technology.html Accessed July 27, 2017.
24. Pedregosa, F., G. Varoquaux, A. Gramfort, V. Michel, B. Thirion, O. Grisel, M. Blondel, P. Prettenhofer, R. Weiss, V. Dubourg, J. Vandelplas, A. Passos, D. Cournapeau, M. Brucher, M. Perrrot, and E. Duchesnay. Scikit-Learn: Machine Learning in Python. *Journal of Machine Learning Research*, Vol. 12, 2011, pp. 2825–2830.

The Standing Committee on Transportation Planning Applications (ADB50) peer-reviewed this paper (18-05791).

Sage Publications USA
sagepub.com

ISBN 978-1-5108-8252-2